Papermaking
for the first time®

Papermaking
for the first time®

Rhonda Rainey

Sterling Publishing Co., Inc.
New York
A Sterling / Chapelle Book

Chapelle:

Jo Packham, Owner

Cathy Sexton, Editor

Staff: Areta Bingham, Kass Burchett, Jill Dahlberg, Marilyn Goff, Holly Hollingsworth, Susan Jorgensen, Barbara Milburn, Linda Orton, Karmen Quinney, Cindy Stoeckl, Kim Taylor, Sara Toliver, Desirée Wybrow

Photography: Kevin Dilley for Hazen Imaging, Inc.
Gallery Photography: Various professional photographers unknown by name, unless indicated.

If you have any questions or comments or would like information on specialty products featured in this book, please contact Chapelle, Ltd., Inc., P.O. Box 9252, Ogden, UT 84409 • (801) 621-2777 • (801) 621-2788 Fax • e-mail: chapelle@chapelleltd.com • website: www.chapelleltd.com

Library of Congress Cataloging-in-Publication Data Available

Rainey, Rhonda.
    Papermaking for the first time / Rhonda Rainey.
        p. cm.
    Includes index.
    ISBN 0-8069-2508-6
    1. Papermaking. 2. Paper, Handmade. I. Title

TS1105.R35 2002
676'.22--dc21   2001049216

10 9 8 7 6 5 4 3 2 1

Published by Sterling Publishing Company, Inc.
387 Park Avenue South, New York, NY 10016
© 2002 by Rhonda Rainey
Distributed in Canada by Sterling Publishing
c/o Canadian Manda Group, One Atlantic Avenue, Suite 105
Toronto, Ontario, Canada M6K 3E7
Distributed in Great Britain and Europe by Cassell PLC
Wellington House, 125 Strand, London WC2R 0BB, England
Distributed in Australia by Capricorn Link (Australia) Pty. Ltd.
P.O. Box 704, Windsor, NSW 2756, Australia
Printed in China
All Rights Reserved

Sterling ISBN 0-8069-2508-6

Table
of Contents

Papermaking for the first time

Introduction

Paper has intrigued and fascinated me for as long as I can remember. As a child, I loved paper dolls, books, and coloring books. My grandmothers entertained me with butcher wrap and brown paper bags saved from the meat market and grocery stores long before anyone had thought of plastic trays, bags, and shrink wrap. By the time I was into my college years, I had determined that watercolors and drawing were my mediums of choice, in part, because I loved working on the variety of surfaces paper offered.

In the early 1990s, a friend and fellow artist, Joanne Crane, introduced me to the wonders of Japanese papermaking. I was smitten! Since that time, I have had the good fortune of being able to take university classes and workshops which have expanded my love and knowledge of handmade papers and its creative possibilities.

As with so many things we take for granted in everyday life, paper came about as a necessity. Ancient civilizations advanced and spread, creating the need for surfaces on which writing and records could be kept and the spoken word transcribed. Marks and symbols were applied to stone, clay, bones, wood, metal, and even leaves. There were many problems with these materials. If they were permanent, they were cumbersome. If they were lightweight and portable, such as palm leaves, bamboo stalks, and cloth, they were somewhat fragile and short lived. The ancient Egyptians learned how to work the papyrus plant into a thin board-like substance and the Persians developed parchment from the treated skins of sheeps and goats.

The invention of true paper is attributed to a Chinese court official by the name of Ts'ai Lun. In the early part of the second century A.D., he discovered that materials such as old fishing nets, rags, hemp waste, and other plant fibers could be soaked and then beaten into single fibers. This pulpy mass was stirred into a vat of water and then lifted on a woven screen which had been stretched across a frame. As it drained and dried, the pulp formed a matted material, the ancestor of today's papers. Those first papers were of poor quality, primarily because of the materials used to make the pulp. There is a saying in the papermaking community that "a paper is only as good as the things that go into it." As papermaking methods were improved and refined, this "discovery" became the favored and most accepted writing surface.

The Chinese closely guarded their papermaking knowledge for several centuries. Some 500 years after Ts'ai Lun's apprentice experimented with his master's ideas, the art of papermaking spread to Korea and then to Japan via Buddhist monks and their sacred texts which contained papers made from mulberry fiber. The Japanese made further improvements by adding local varieties of plant matter and by recycling paper which had been previously used into new sheets.

In the Asian world, paper was used as a sacred and ceremonial material. As its aesthetic and functional uses increased it became an inseparable part of daily life.

In 751 A.D., the "secret" of the Far East became a part of world knowledge when a group of warring Arabs took Chinese prisoners during a battle at Samarkand, a city in central Asia, located north of today's Afghanistan. The prisoners were forced to divulge their knowledge and working methods to the Arabs, who carried the information throughout the Mediterranean region, east into India, and finally west into Italy and Spain. Paper's journey from China to Europe had taken one thousand years.

How to use this book

The purpose of this book is to provide an understanding of the techniques and tools used to produce unique, personal handmade papers from waste or scrap paper. The excitement and pleasure of creating something new from resources as ordinary and abundant as scrap paper feels like nothing short of magic.

There are no long cooking processes and each step can be safely accomplished in the kitchen. The approach is simple and the rewards are immediate. Best of all, there are few rules and only suggested recipes. As with all artistic pursuits, imagination and experimentation go a long way toward achieving successful results.

The techniques are explained step by step and, where necessary, photographs and diagrams are provided for clarity. Papermaking techniques and projects are organized in order of difficulty. Those projects which are most difficult involve more steps and may include several papermaking and crafting techniques combined. To begin, choose projects for your level of ability and then expand.

The following suggestions should give an idea as to how much scrap paper is needed to complete your projects with several sheets of handmade paper left over. Paper that is torn soaks up more water than paper that is cut, making less work for your blender. Always remove staples and cellophane in window envelopes before tearing and measuring. All dry measurements should be tightly packed. Four sheets of 20–24 lb. (standard weight) 8$\frac{1}{2}$" x 11" paper is equivalent to one cup of torn paper. Twelve sheets of standard-weight paper will yield approximately nine sheets of 8$\frac{1}{2}$" x 11" handmade paper. One sheet of wrapping tissue, 20" x 30", will render one cup dry, crushed paper. This will make three 8$\frac{1}{2}$" x 11" sheets of thin paper. The amount of paper you make will be approximately 75% of the amount of paper torn up.

The term "recyclable" is used throughout this book. This refers to waste or scrap paper—that which is to be torn, soaked, and processed for handmade sheets. Industry has already created this paper by adding sizing, optical brightners, fillers, colorants, and substances which make the sheet opaque. This pulp, with all of its additives, is called "furnish." Many qualities of the industrial furnish are maintained in the handmade sheet. Recycled paper is that which has been paper at one time and has now been reprocessed and reformed into new sheets.

Papermaking does not require long hours of practice, however, as you work, you will begin to develop a "feel" for what you are doing and a natural rhythm. Most papermakers enjoy the process as much as the final result.

This book is lovingly dedicated to my parents, Oreon and Alda Brown.

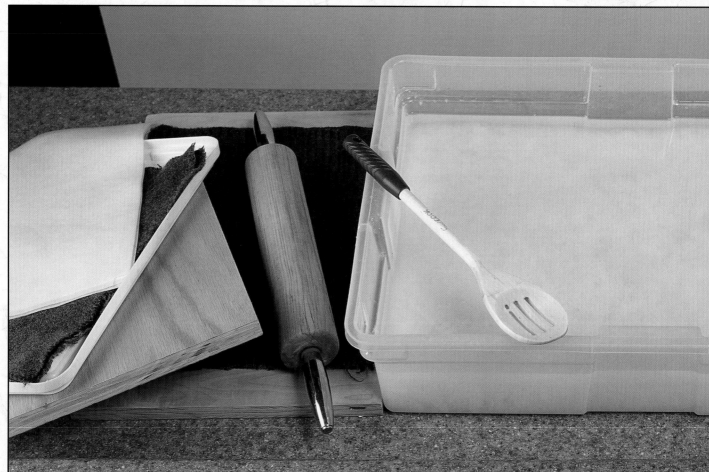

Section 1:
papermaking basics

What materials and tools do I need to get started?

1 Abaca (not shown)

Abaca is a traditional papermaking fiber that can be purchased preprocessed and pressed into a heavy-weight sheet. Pieces are torn from the sheet, soaked, and added to recycled paper pulp for added strength. Abaca can be ordered from papermaking suppliers.

2 Blender

An electric blender is used to break down soaked paper fibers into a pulpy mass.

3 Bowl (not shown)

A large bowl, plastic or metal, is used for soaking paper before pulping, and for holding waste pulp and water for disposal.

4 Clear Glass Jar with Lid

A small clear glass jar with a tight-fitting lid is used to check the consistency of the paper pulp as it comes from the blender.

5 Cotton Linter (not shown)

Cotton linter, made from 100% cotton fiber, is a traditional papermaking fiber that can be used alone or added to recycled pulp for more strength. It is pure white and takes dyes and colors well. Cotton linter generally can be found in arts and crafts stores or from papermaking suppliers.

6 Couching Cloths
(pronounced kooch-ing)

Eight to twelve couching cloths will be needed. They are used to separate wet sheets of paper as they are transferred from the mold to the couching mound. Couching cloths can be made from dressmaker's heavy-weight inter-facing fabric. It must be the type that is sewn in, not the iron-on type. Couching cloths should be the same size as the felts. They can be washed in cold water and reused.

7 Dish Towels—Cotton and Terry Cloth

A cotton dish towel or piece of muslin is used as a pressing cloth over the paper as it is ironed. This prevents "glazed" marks that can sometimes occur. Terry cloth dish towels are used to absorb moisture as the paper is rolled with a rolling pin.

8 Drying Boards

A nonabsorbent drying board such as a piece of Plexiglas® or a Formica®-covered board or countertop works well for drying sheets of damp paper once they have been pressed. These hard, smooth surfaces allow the paper to dry flat with no warping or wrinkling.

9 Felts

Two pieces of felt are needed. The pieces of blanket-like fabric, traditionally wool, act as a support for the wet papers as they are couched. Wet felts before using. One is placed over the bottom pressing board and the other on top of the mound of paper next to the top pressing board. They should be cut approximately 4" larger in width and length than the mold and deckle. Sometimes wool blankets can be found at an Army surplus store or at a secondhand store. They are generally inexpensive if they are seconds or damaged. Launder with cold water, soap, and disinfectant. Let dry, then trim to size and bind edges. Felts are also referred to as pressing blankets.

10 Iron and Ironing Board (not shown)

An iron is used to dry and flatten damp paper. It is also necessary if spray laundry starch is used to size the sheet.

11 Measuring Cup with Pouring Spout

The measuring cup can be glass or plastic and is used to measure wet paper and to pour pulp when using the poured method.

12 Mold and Deckle

The mold is considered to be the most important piece of papermaker's equipment. It is a wooden frame, covered with a piece of mesh or fiberglass window screen. The wet mold is lowered, mesh side up, into the vat of water and pulp, lifted out, and allowed to drain. As the water drains through the screen, the paper fibers settle onto the mesh and form a sheet of paper.

The deckle is an open wooden frame exactly the same size as the mold. It fits on the mold to help contain the pulp and shape the sides of the paper. It is lifted and set aside before the wet sheet of paper is removed from the mold. Sometimes a deckle is not used. Instead, the paper pulp is allowed to fall over the edges of the mold, forming the beautiful "feathered" edges that are characteristic of a handmade sheet.

13 Needle and Thread (not shown)

A needle is helpful in removing foreign matter from a dried sheet. Use it as if removing a splinter or sliver. It is easier to find and handle if threaded.

14 Paintbrush

A 3" flat paintbrush is used to brush the wet paper onto the drying boards.

15 Plastic Jug with Handle (not shown)

A gallon-sized plastic jug with a handle is used to add water to the blender and to the vat. Various sized plastic jugs are used to hold paper pulp after it has been blended and before it is added to the vat.

16 Plastic Sheeting (not shown)

Plastic sheeting or a vinyl tablecloth is used to cover and protect work surfaces.

17 Plastic Tray (not shown)

A plastic tray or a rustproof cookie sheet with sides holds the damp felts and wet couching cloths.

18 Pressing Boards

Two pressing boards will be needed. Pressing boards are used to protect the mound of paper as the water is pressed from it. They should be slightly larger than the mold and deckle, rigid, and waterproof. If using wood boards, they must be sealed with two coats of polyurethane or acrylic varnish before being used for the first time—make certain to seal the edges.

19 Rolling Pin

A rolling pin is used to continue the pressing process.

20 Sealable Plastic Bags (not shown)

Sealable plastic bags are used for storage.

21 Semiprocessed Pulp (not shown)

These fibers, generally cotton and abaca, are partially processed and pressed into heavy sheets that can be reconstituted in water and made into sheets of paper. Added to recycled paper pulp, they make papers of strength, quality, and long life.

22 Sizing

Sizing is an important element to handmade papers because it enhances the integrity of the sheet. A good sheet of watercolor paper will be sized both internally and externally.

Unflavored gelatin is used as internal sizing added to the pulp in the vat during the paper

forming stage. It may also be used externally where the paper is dipped into it.

Spray starch is used as external sizing and is sprayed onto both sides of a sheet of paper before being pressed into the sheet with a warm iron.

Cellulose powder is used to size papers internally and externally and is available at wallpaper stores. A little goes a very long way.

23 Sponges

Cellulose sponges are used to remove excess water from the mold when couching the sheet of paper and for cleanup. The same sponge should not be used for both processes.

24 Spray Bottle (not shown)

A fine-mist spray bottle filled with water is used to mist the mold and deckle to keep them damp and to add moisture to the pressing cloth.

25 Strainer

A strainer is used to catch leftover pulp, which should NEVER be run down a drain. It will cause a serious clog. Small amounts of leftover pulp should be disposed of in the trash. Larger amounts can be strained, put into a sealable plastic bag, and kept in the refrigerator (or frozen) for later use. It is advisable to catch the waste water and put it down the toilet.

26 Tweezers (not shown)

Pointed tweezers are used to remove fine, foreign matter that may find its way into the vat of pulp or onto a wet sheet of paper. These may include pet hair or a piece of waste paper that was not fully processed.

27 Vat

A large plastic leakproof tub is needed to contain the paper pulp and water. A six- to seven-gallon plastic storage container, with a depth of six inches, is ideal because it is large enough to accommodate an 8½" x 11" mold and deckle. It also can be used as storage for your tools when finished. For a smaller mold and deckle, 5" x 7" or 6" x 8", a plastic dishpan works well.

28 Waste Paper (not shown)

Waste paper is paper for recycling—used as a base material for all projects presented in this book. Almost any kind of paper can be recycled, but some are more suitable than others. Among the best are computer papers, photocopier paper, envelopes, printed programs, uncoated butcher wrap, bank checks, writing paper, and brown wrapping paper (if it has not already been recycled).

Heavily printed paper or highly acidic paper, such as newsprint, can present problems. Newspaper will become yellow and brittle in a very short time. It can be used in combination with other fibers, but as a decorative element only. Some inks will leave the paper pulp a murky gray and make an annoying film on your papermaking equipment.

The quality of the handmade paper you produce will be determined by the quality of the waste paper used to make it.

Waste paper and paper for recycling are interchangeable terms.

29 Water (not shown)

A supply of clean, fresh water is necessary for every step of the paper forming process.

30 Weights (not shown)

Weights are used to flatten damp paper after it has been pressed and occasionally before it is used if a curl or cockle appears in the sheet. Heavy books, bricks, or a bucket of sand may be used for this purpose. When using books or magazines, it is a good idea to place them in a plastic bag before using to help protect them from becoming wet.

31 Wooden Spoon

A wooden spoon is used to stir the vat each time pulp is added and each time a sheet of paper is made.

32 Work Surface (not shown)

A table large enough to hold the vat, felts, and couching cloths will be needed. A kitchen counter is generally too high to work from comfortably.

How do I care for my equipment?

Taking care of your papermaking equipment will considerably extend its life and usefulness.

The blender is the most expensive piece of equipment for the papermaker; and the most important thing to remember while using it is not to overload it while processing pulp. Using $1/3$ to $1/2$ cup packed, torn, and soaked paper to three cups water is the maximum amount that should be processed at one time. If a heavier pulp is needed to make a thicker sheet of paper, drain some of the water from the pulp after blending it, but before pouring it into the vat.

After using the blender, unplug it from the wall recepticle. The blender container and blades should be washed and dried thoroughly and the lid taken apart and washed. This is the best way to insure that unwanted, stray pulp fiber will not show up in your next project. Don't forget the area around the buttons. Cotton swabs are handy for cleaning in these "tight" places.

Wipe up spills as they occur so there is not excessive moisture around the blender or puddles on the floor.

Couching cloths and felts should be hand-washed in cool water. After rinsing and wiping out the vat, fill half-way with clean, cool water and swish cloths two or three at a time, then hang cloths to dry. Several lightweight cloths may be hung together for drying. Wooden dowels placed across a bathtub are an inexpensive and easy way to make a drying rack.

Change the water in the vat and repeat process for felts. However, because of their weight and the time it takes for them to dry, it is best to hang these individually to dry.

Plastic jugs and containers should be rinsed and dried, and drying and pressing boards wiped down.

The mold and deckle should be rinsed and stubborn pulp fibers removed. If necessary, a brush can be used for this purpose. They should be air-dried standing up as this will help prevent warping.

If a large amount of pulp (two quarts or more) is left over from a project and is to be kept for future use, strain it. Shake the excess water from it while it is still in the strainer, then put it into a plastic zipper-type food storage bag. Do not squeeze the pulp or it will "knot" up, making it unsuitable for sheet paper use. The storage bags can then be frozen. Make certain to label the bags with their contents.

Any other leftover pulp should be disposed of in the trash or put down the toilet. Waste water should also be flushed down the toilet, and never be poured down the drain.

Even though many items used as tools for making paper can be found in the kitchen, it is advisable not to reuse them for cooking purposes once they have been used for paper-making.

Most equipment, after it is dry, can be stored in the vat. Any creative endeavor is easier and more successful when supplies and tools are organized and accessible.

How do I make a mold and deckle?

What You Need To Get Started:

Brads
Craft knife
Duct tape
Fiberglass
   window screen
Hammer
Heavy-duty scissors
L-shaped brackets
   and screws
Paintbrush,
   1/2" flat
Paper towel
Pencil
Push pins
Ruler
Sandpaper,
   fine-grit
Screwdriver
Staple gun
   and staples
Water-based
   satin varnish
Waterproof
   wood glue
Weights
Wood strips,
   1" x 1" x 8 1/2" (4)
Wood strips,
   1" x 1" x 12 1/2" (4)
Work surface

Note: The inside dimensions of the mold and deckle frames determine the size of the finished sheet of paper.

Here's How:

1. Sand raw edges of wood strips, if necessary.

2. Make two identical frames by placing two of the 8 1/2" wood strips perpendicular to two of the 12 1/2" strips. The longer sections should sit to the outside of the shorter sections.

3. Apply waterproof wood glue at the frame joints, following manufacturer's instructions. Let dry.

4. Stand the frame on its side. Tapping downward with a hammer, center and drive a brad through each joint into the frame.

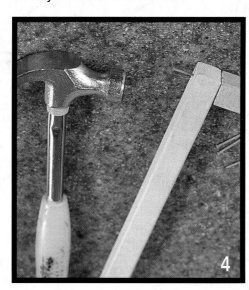

5. Sand edges and rough areas until smooth. Wipe away excess dust with a paper towel.

6. Using a flat paintbrush, apply three coats of varnish to all areas of both frames. Let dry after each application.

7. Set one frame aside—this is the finished deckle.

8. Using a ruler and pencil, mark and rule a stapling guideline 1/4" from the opening of the mold.

9. Using heavy-duty scissors, cut one piece of screen 4" larger than the mold.

10. Lay the mold flat on the work surface and center the screen on top.

11. Using push pins, temporarily secure screen to mold at each corner.

12. Using a staple gun, attach the screen at guideline on one long side of mold. Leave 1/4" between each staple. Remove push pins.

13. Tightly pull the screen across the mold at center of opposite side. Staple from center to corners while continuing to pull screen as tightly as possible.

14. Repeat Steps 12 and 13 across both short sides.

15. Using the ruler and a craft knife, trim the screen just short of the mold's outside edge.

16. Using duct tape, cover stapled edges and edge of screen. Carry tape over outside edge of mold and trim flush with bottom of mold.

17. Using a screwdriver, attach the L-shaped brackets to reinforce the corners.

Section 2:
basic techniques

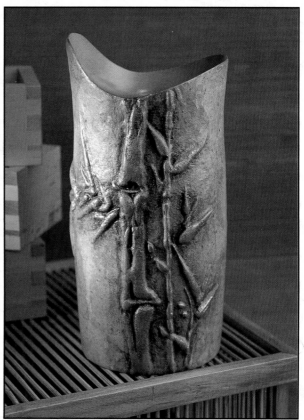

1
technique

Blender
Bowl, large
Clear glass jar with lid
Couching cloths
Dish towels,
    cotton and terry cloth
Drying boards
Felts (2)
Funnel
Iron and ironing board
Measuring cup with
    pouring spout
Mold and deckle
Needle and thread
Paintbrush,
    3" flat
Plastic jug with handle
Plastic sheeting
Plastic tray
Pressing boards
Rolling pin
Sealable plastic bags
Sponges (2)
Spray bottle, fine-mist
Strainer
Tweezers
Vat
Waste paper
Water
Weights
Wooden spoon
Work surface

How do I make paper pulp and form sheets of paper using the dipping method?

When making paper and dipping a mold and deckle into a vat of pulp to pull a sheet of paper, the technique is referred to as the Western or European style. This method works well for making paper from shorter fibers and is particularly well suited to use with recyclable papers. One advantage to using this method is that many sheets can be made in a short period of time.

Dipped Sheets of Paper

Here's How:

Tearing and Soaking Waste Paper

1. Remove any foreign matter such as cellophane, staples, or glue residue from waste paper.

2. Tear waste paper into pieces approximately 1" square. Put squares into a large bowl. Cover with water and soak overnight.

Note: Tissue papers and napkins can be successfully soaked in 30 minutes. Very heavy paper, such as watercolor paper, should be soaked for two to three days.

Blending/Beating

3. Measure $1/3$ to $1/2$ cup torn paper and add to three cups water in blender container.

4. Using one of the pulse buttons, process paper in short bursts for 15 seconds.

5. Turn blender to high and blend for another 10 seconds.

Note: If the blender sounds like it is laboring or feels sluggish, add more water.

6. Stop blender and check fiber consistency. Fiber may be tested by putting 1 teaspoon pulp into a jar of water. Seal with lid and shake. If pulp shows clumps or "flecks" of whole paper, more processing is needed. Pulp should have a smooth consistency.

Note: Exact timing is not critical to this process, but overblending will break fibers down too much, causing a weak paper.

7. Continue processing soaked paper until you have the needed amount plus two blenders surplus. Pour the pulp into a plastic jug(s) to store until ready to use.

Note: Refrigerate leftover pulp unless it is to be used the same day.

Preparing the Work Surface and Vat

8. Set up work surface with pressing boards, wet felts and couching cloths, sponges, tweezers, and spray bottle to mist the mold and deckle before forming sheets.

9. Pour pulp from plastic jug into the vat until half full. Add two quarts water.

Note: The remaining pulp is used to replenish the vat as sheets of paper are pulled, ensuring

paper of a more consistent thickness. This should be done approximately every three sheets.

10. Using a wooden spoon, stir contents of the vat until pulp is evenly dispersed in water.

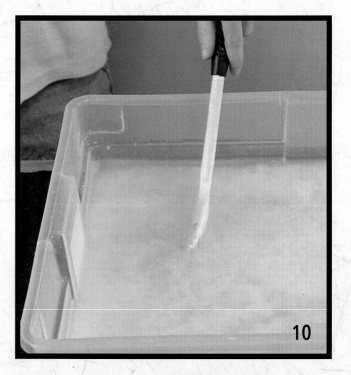

Dipping Sheets of Paper

11. Using a spray bottle, lightly mist the mold and deckle with water before using.

12. Working quickly, before stirred pulp settles to bottom of vat, place deckle over mold next to mesh.

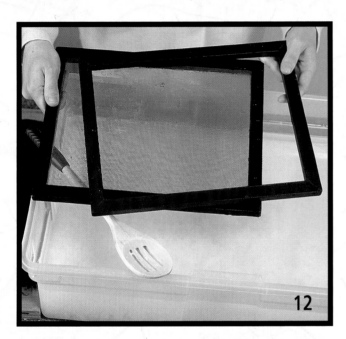

13. Holding both pieces firmly together at the short sides and keeping the mold vertical to the vat, dip into pulpy water at far side of vat.

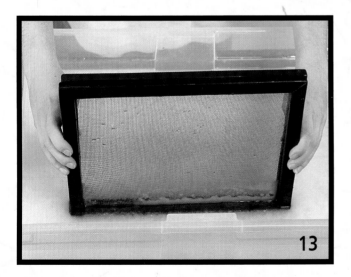

14. Using a smooth, continuous motion, tilt mold and deckle until they are horizontal. Pull toward front of vat until both mold and deckle are completely immersed.

15. Keeping mold level, lift straight up out of water.

16. With mold still in horizontal position, and while pulp is very watery, give a quick shake from front to back and from side to side in a sifting motion. This is called the "papermaker's shake." It helps even out the pulp and bond the fibers for a stronger paper.

17. Holding the mold and deckle over the vat and slightly tilting, drain excess water.

18. When water has stopped draining, place the mold on a flat work surface and remove the deckle, making certain no water drops fall onto the screen and pulp. Drips will cause holes called "papermaker's tears."

Note: If sheet is unsatisfactory, it may be returned to the vat and reused simply by turning the mold over and touching it to the water and pulp in the vat. The sheet will slide off. Remist the mold and deckle. Stir pulp in vat and try again.

Couching

19. On a flat work surface, prepare the couching mound using one pressing board as a base, followed by a wet felt, then a wet couching cloth. Smooth out any wrinkles.

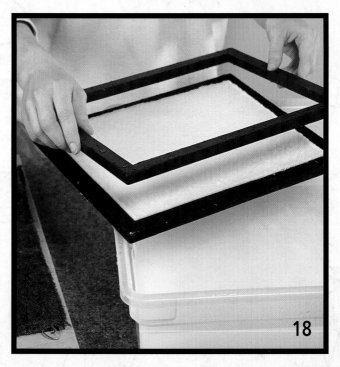

20. Rest long edge of frame along edge of couching cloth. Gently tip mold and pulp face down.

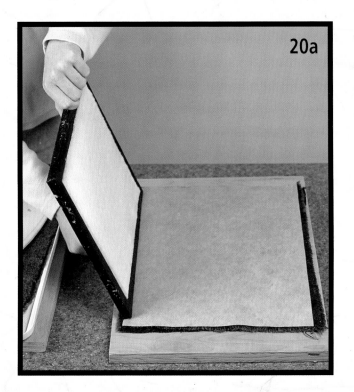

21. Using a sponge, absorb the excess moisture through the screen from the back of the mold.

Note: Heavier papers generally roll off the mold with no difficulty. Thinner papers may require sponging.

22. Using a rolling motion, lift the mold from the couching cloth.

23. Cover the sheet of paper with another damp couching cloth, making certain not to drag it across the paper.

24. Repeat Steps 11–23 for each additional sheet, making certain each sheet of paper rests neatly and as evenly on top of the previous sheet (separated by the couching cloth) as possible.

Note: This stack of wet paper is called a post and may consist of 10 to 15 sheets, depending on the number of couching cloths available.

25.   Complete the post by placing a wet felt and a second pressing board on the top.

Pressing to Remove Water

26.   Cover the floor with plastic sheeting.

27.   Place pressing boards and paper on plastic.

28.   Step onto the top pressing board and, gently moving around, press all areas.

29.   Place weights on top of the pressing boards and let set for one hour.

Drying

30. Remove the weights and lift the post to the tabletop. Remove the top pressing board and felt. Set aside.

31. Place drying boards on a flat surface.

32. Lift the first wet sheet and couching cloth together from the mound. Hold by one corner of the cloth and pick up at diagonal corner with paper on underside of the cloth.

33. Place paper (face down) and couching cloth on the drying board.

32–33

34. Using a flat paintbrush, brush over couching cloth and paper.

34

35. Lay a terry cloth dish towel over the couching cloth and paper.

36. Using a rolling pin, gently roll back and forth across the dish towel and couching cloth.

36

37. Remove the dish towel.

38. Gently lift couching cloth from pressed paper.

39. Repeat Steps 32–38, continue to separate sheets. Let dry.

Note: Check dryness by placing wrist pulse point to paper surface. If paper feels cool to the touch, it still has moisture in it. Let it continue to dry. If paper is removed from drying surface before the sheet is dry, it will curl or cockle. Temperature and humidity will affect drying times.

Ironing to Smooth and Flatten

Note: Sheets of paper that are board-dried will show two distinct surfaces. The upper surface of the paper will be softer to the touch and slightly rough. The surface which has dried next to the board will feel harder and look much smoother. Ironing paper after it has dried will help even out both surfaces and also strengthens the sheet of paper.

40. Using a spray bottle, lightly mist paper on the front and the back with water.

41. Place damp sheet between two cotton dish towels and press with an iron set at medium heat.

Weighting to Smooth and Flatten

Note: Heavier sheets of paper that have curled or cockled during the drying process must be flattened by this method.

42. Using a spray bottle, lightly mist paper on the front and the back with water.

43. Place damp sheet on hard flat surface. Using fingers and palms of hands, gently stretch paper from center of sheet to edges, making certain not to tear the paper.

44. Repeat Steps 42–43 as necessary.

45. Place sheets of handmade paper between sheets of clean dry paper. Stack, then weight with heavy books, boards, or bricks.

Note: Do not place damp sheets of paper in a book or the book's pages will curl.

Design Tip: Journals become even more personal and artistic when they are created from handmade papers. When using markers or liquid ink, a heavier gelatin sizing is required to seal the surface of the sheet. This prevents the ink from feathering or bleeding through the papers. If writing with a ball-point pen or a gel-type ink, lightly sizing the paper with spray starch is adequate.

How do I use paper pulp to form sheets of paper using the pouring method?

According to many scholars, the papermaking style called "pouring" is the oldest of the sheet-forming methods. It is ideal if you have only a small amount of pulp to use or if you want to make a very large or thick sheet. Mixing various colors and types of pulp also works well with this method. Allowed to dry on the screen, the paper will take on a heavily wrinkled and unique texture.

2
technique

**What You Need
To Get Started:**

Blender
Bowl, large
Clear glass jar with lid
Drying boards
Funnel
Measuring cup with
    pouring spout
Mold and deckle
Needle and thread
Plastic jug with handle
Sealable plastic bags
Sponge
Strainer
Tweezers
Vat
Waste paper
Water
Wooden spoon
Work surface

Poured Sheets of Paper

Here's How:

Making Paper Pulp

1. Blend pulp from waste paper and pour into a plastic jug. Refer to Technique 1: Tearing and Soaking Waste Paper and Blending/Beating on pages 21 and 22.

Pouring Sheets of Paper

2. Set up a mold. Place mold over vat to catch water from pouring process. Mold must be level. If necessary, prop at corners of mold.

3. Pour thin pulp into measuring cup.

Note: If pulp is too thick it will pour in clumps. Thin with water.

4. Pour pulp from measuring cup onto screen. Begin at outside edges of mold and slowly pour widthwise. Fill in entire screen area, refilling measuring cup as necessary.

Note: For stronger papers, pour pulp lengthwise onto screen. Pour next layer widthwise. Repeat as necessary. For thinner sheets, pulp should be poured to $1/8$"—for heavier sheets, pour to $1/4$".

5. After pulp has stopped dripping, move mold to level area and let dry for several days.

Note: Temperature and humidity will affect drying times.

6. Remove paper from the mold by lifting a corner, working around the edges, and to the center.

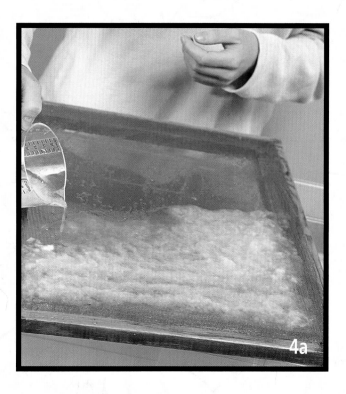

How do I add color to my handmade papers using colored waste paper?

Coloring handmade papers can create interesting effects in a broad range of colors. The simplest and most direct approach is to recycle colored paper. Paper napkins, wrapping tissue, colored papers, and scrapbooking papers are easily available. Some papers printed with colored inks also work well. However, the final effect with these is much more subdued than it is with papers dyed in the pulp. To determine whether a sheet has been dyed or printed, tear the sheet. If the torn edge is the same color as the sheet, the sheet was dyed. If the torn edge is white, the sheet was printed.

The recycling process is the same for colored papers as it is for white paper—tear, soak, and blend. The colored pulp may be used full strength or blended with white or other colors. If an evenly colored paper is desired, the pulps must be mixed in the blender. If a speckled paper is desired, different colors of pulp may be mixed in the vat.

What You Need To Get Started:

Supplies listed for Technique 2 on page 30

Craft knife
Découpage medium, matte finish
Double-tipped stylus
Mold, 18" x 24"
Paintbrush, 1½" china bristle
Plastic cup
Plastic jugs with handles
Straightedge
Tape
Waste papers to yield one 18" x 24" sheet:
  For background:
    Dark brown, 12 sheets
  For leaves:
    Light green, 3 sheets
    Olive green, 3 sheets
  For roses:
    Dark coral, 4 sheets
    Light coral, 4 sheets
    Medium coral, 4 sheets
Wire-edged ribbon

"Here's How:" begins on page 34.

Folio Cover

Here's How:

Making Paper Pulp

1. Blend thin pulp from each of three colors of waste papers for the roses and pour into separate plastic jugs. Refer to Technique 1: Tearing and Soaking Waste Paper and Blending/Beating on pages 21 and 22. Repeat for each of two colors of waste papers for the leaves and one color for the background.

Note: Wet pulp will always appear more intensely colored and darker than the paper made from it.

Caution: Many of the coloring agents used in paper will stain, even in the recycling process. Take care to wipe up spills immediately. Wash felts and couching cloths after use with colored pulps.

Pouring Sheets of Paper

2. Set up an 18" x 24" mold. Place mold over vat to catch water from pouring process. Mold must be level. If necessary, prop at corners of mold.

3. Pour thin pulp into measuring cup.

Note: If pulp is too thick it will pour in clumps. Thin with water.

4. Beginning with the lightest shade for the rose, pour a 4" spiral. Randomly repeat pattern, leaving spaces for leaves. Continue pouring roses in spiral pattern using colored pulps.

5. Pour leaves in a triangular pattern, joining each leaf to outside edges of roses.

6. Beginning at the outside edges of mold, slowly pour background color onto screen to fill in all areas around and between roses and leaves. Fill in entire screen area, refilling measuring cup as necessary.

Note: For stronger papers, pour pulp lengthwise onto screen. Pour next layer widthwise. Repeat as necessary. For thinner sheets, pulp should be poured to $1/8$"—for heavier sheets, pour to $1/4$".

7. After pulp has stopped dripping, move mold to level area and let dry for several days.

Note: Temperature and humidity will affect drying times.

8. In a plastic cup, mix one part découpage medium to one part water. Stir with the handle of the paintbrush.

9. Using a china bristle paintbrush, apply the diluted découpage mixture over surface of paper, while it is still on the mold. Let dry.

Caution: Do not use too much découpage mixture or the paper will not come loose from the screen.

10. Remove paper from the mold and repeat Step 9 on reverse side.

Making Folio Cover

11. Enlarge Folio Pattern 310% on page 35. Tape pattern to back side of paper.

12. Using a straightedge and a craft knife, cut folio from paper.

13. Using the straightedge and the large end of a stylus, gently score paper where dotted lines appear on pattern. Fold side flaps to center, bottom flap up, and top flap down.

14. Tie with wire-edged ribbon.

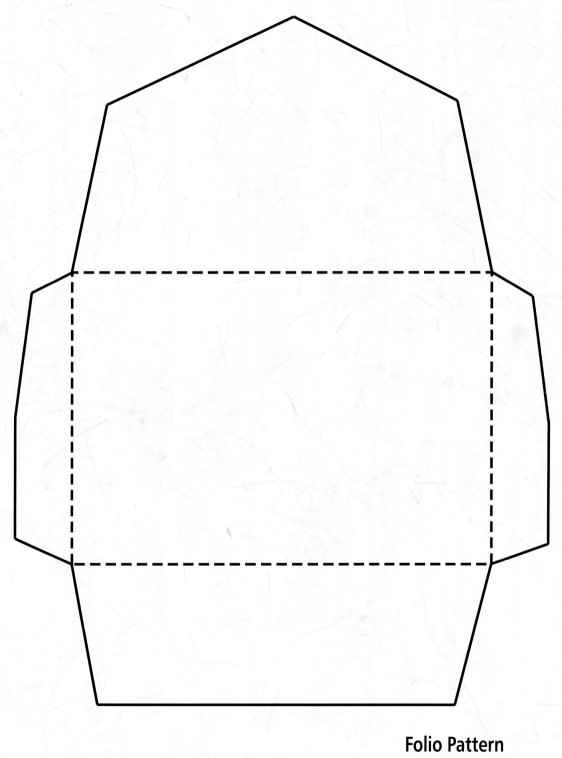

Folio Pattern

4

technique

What You Need To Get Started:

Supplies listed for Technique 1 on page 20

Gelatin, unflavored
Measuring cup with pouring spout
Measuring spoons
Paintbrushes
Plastic container, approximately 10" x 13"
Waste paper to yield eight 8¹/₂" x 11" sheets: White, 24 sheets
Water-based paints

How do I add sizing to my handmade papers for watercolor or other wet medium use?

Sizing is a process which seals and strengthens paper surfaces. In order to use wet mediums, such as ink or watercolors, a glue-like substance must be added to the paper to make it less absorbent. Unsized paper, called watermark, is much like blotter paper. Inks and liquid paints will bleed and feather and may soak through.

Gelatin sizing is traditionally used in the vat (internal), or it can be used after the paper has been formed (external). It is suggested that paper to be gelatin-sized be made at least three weeks in advance of its use. This gives the sheet time to age, which helps strengthen and settle the fibers. In addition, there is less chance that the paper will come apart as it is being handled.

Greeting Card

Here's How:

Making Paper Pulp and Sheets of Paper

1. Blend pulp from waste paper and pour into a plastic jug. Refer to Technique 1: Tearing and Soaking Waste Paper and Blending/Beating on pages 21 and 22.

2. Finish making the paper. Refer to Technique 1: Preparing the Work Surface and Vat on page 22, Dipping Sheets of Paper on pages 23 and 24, Couching on pages 24–27, Pressing to Remove Water on page 27, and Drying on pages 28 and 29.

Adding Sizing

3. Dissolve 1¹/₄ teaspoons gelatin in one cup boiling water. Pour into the plastic container and add warm water until mixture is 1" deep.

4. Gently submerge each sheet of paper for 30 seconds. Using the handle of the wooden spoon, carefully lift each sheet from the plastic container and transfer to a drying board. Let dry.

5. Using a spray bottle, lightly mist paper on the front and the back with water.

6. Cover damp sized paper with a cotton dish towel and press with iron on medium heat.

Caution: Make certain to continually move the iron. Avoid leaving it in one position for too long as paper scorches as easily as fabric.

Making Greeting Card

7. Fold the sheet in half and paint any design on the front with water-based paints.

Menu

Tuscan Garlic Bread with
Tomatoes, Basil and Mozzarella Cheese
Ceasar Salad
Grilled Breast of Chicken
Served over Bow Tie Pasta with Lemon and Herbs
Steamed Vegetables and Red New Potatoes

Wines

Hogue 1997,   Columbia Valley
Sauvighon Blanc

Fetzer 1996,   Eagle Peak
Mendocino Country
Merlot

How do I add a light sizing to my handmade papers for computer printer use?

There are a number of substances and methods used to size handmade papers—in this case, spray starch. Recipes and amounts vary according to the paper's intended use and the type of pulp from which it is made. Some experimentation may be necessary. As you work, it is helpful to keep a notebook of successes and difficulties.

What You Need To Get Started:

Supplies listed for Technique 1 on page 20

Color printer
Computer
Computer software
Spray starch, heavy-duty
Waste paper to yield twelve 8½" x 11" sheets:
White, 16 sheets

Party Menu

Here's How:

Making Paper Pulp and Sheets of Paper

1. Blend pulp from waste paper and pour into a plastic jug. Refer to Technique 1: Tearing and Soaking Waste Paper and Blending/ Beating on pages 21 and 22.

2. Finish making the paper. Refer to Technique 1: Preparing the Work Surface and Vat on page 22, Dipping Sheets of Paper on pages 23 and 24, Couching on pages 24– 27, Pressing to Remove Water on page 27, and Drying on pages 28 and 29.

Adding Sizing

3. Using a spray bottle, lightly mist paper on the front and the back with water until both sides are uniformly damp.

4. Place damp paper on cotton dish towel on ironing board and liberally spray both sides with spray starch.

5. Cover damp sized paper with another cotton dish towel and press with iron on medium heat.

Caution: Make certain to continually move the iron. Avoid leaving it in one position for too long as paper scorches as easily as fabric.

6. Carefully inspect the edges of each sheet and remove any loose or feathered particles.

7. Place one sheet at a time in your printer and print as desired.

How do I tear handmade papers?

Artists and handmade books have always had their own audience. They have gained great popularity the past few years. Many book artists began their careers as papermakers—no doubt looking for more ways to use their handmade papers. The qualities of a beautiful paper include its natural softness and feathered edges.

Accordion Book

Here's How:

Making Paper Pulp and Sheets of Paper

1. Blend pulp from waste papers and embroidery floss and pour into a plastic jug. Refer to Technique 1: Tearing and Soaking Waste Paper and Blending/Beating on pages 21 and 22.

2. Finish making the paper. Refer to Technique 1: Preparing the Work Surface and Vat on page 22, Dipping Sheets of Paper on pages 23 and 24, Couching on pages 24–27, Pressing to Remove Water on page 27, and Drying on pages 28 and 29.

Tearing the Paper and Using as Photo Matting

3. Using a straightedge and a liner paintbrush loaded with water, rule a line as desired. Reload paintbrush as necessary.

4. Gently pull paper apart at the "water line."

5. Using a glue stick, adhere torn paper to book pages. Glue photographs onto top of paper so the torn edges become the matting. Journaling can be added to the blank pages of the book, using an archival-quality marker.

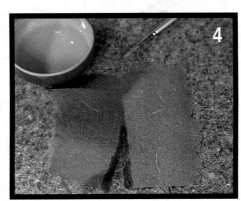

What You Need To Get Started:

Supplies listed for Technique 1 on page 20

Accordion-folded book with blank pages
Archival-quality glue stick
Archival-quality marker
Paintbrush, #6 liner
Photographs
Straightedge
Waste papers and inclusions to yield four 8½" x 11" sheets:
   Dark green, 6 sheets
   Dark green tissue, 1 sheet
   Dark green embroidery floss, 1 yard cut into ½" lengths, then separated

How do I use natural fibers to strengthen my handmade papers?

Each time paper fibers are recycled, they become shorter and weaker. Fortunately, for the papermaker, sheets of preprocessed plant fibers are available from papermaking suppliers and from many craft stores.

Cotton linter is made from the seed hairs collected from cotton plants after they have been ginned for the textile industry. Used alone it makes a paper which is opaque and has a soft feel to it. It is generally a bright white and takes color well.

Abaca fiber comes from the leafstalks of a South American banana plant. The fiber is long, silky, and light beige in color.

Mixed with other pulps, either of these fibers make the paper stronger and easier to work with. Each can be blended with recycled-paper pulp. Cotton linters and abaca sheets are generally sold as heavy sheets—almost like blotter paper. They are easier to tear if soaked first. One hour is sufficient for cotton, and five to ten minutes for abaca.

What You Need To Get Started:

Supplies listed for Technique 1 on page 20

Abaca,
  preprocessed,
  2" x 8" strip
Brass embossing
  stencil, Celtic knot
Craft glue
Craft knife
Double-tipped stylus
Folding bone
Masking tape
Plastic jugs with handles
Straightedge
Waste papers to yield
twelve 8½" x 11" sheets:
  For body of card:
    Dark gray,
      4 sheets
    Medium brown,
      4 sheets
    White, 8 sheets
Waste paper to yield
two 8½" x 11" sheets:
  For dry-embossed
  card insert:
    Light tan, 4 sheets

"Here's How:" begins on page 44.

43

Dry-embossed Card

Here's How:

**Making Paper Pulp
and Sheets of Paper for Body of Card**

1.   Soak abaca strip in a large bowl of water for 10 minutes.

2.   Tear wet abaca strip into sixteen 1" squares.

3.   Blend pulp from three squares abaca and three cups water until mixture is the consistency of oatmeal. Equally divide blended pulp into two plastic jugs. To make pulp for body of card, blend waste papers and the abaca pulp from one of the jugs, then pour into a plastic jug. Refer to Technique 1: Tearing and Soaking Waste Paper and Blending/Beating on pages 21 and 22.

4.   Finish making the paper. Refer to Technique 1: Preparing the Work Surface and Vat on page 22, Dipping Sheets of Paper on pages 23 and 24, Couching on pages 24–27, Pressing to Remove Water on page 27, and Drying on pages 28 and 29.

Note: Make certain to thoroughly rinse the vat before making next batch of paper.

**Making Sheet of Paper
for Dry-embossed Card Insert**

5.   To make pulp for dry-embossed card insert, blend waste paper and the abaca pulp from the remaining jug, then pour into a plastic jug.

6.   Finish making the paper.

Dry-emboss Card Insert

7.   Using a craft knife, trim paper to fit size of stencil design plus 1" on each side.

8.   Secure stencil to paper with masking tape. Turn paper over so stencil is underneath.

9.   Using the end of a folding bone and beginning at the center of the sheet of paper, press firmly and rub over paper's surface.

Note: Design will faintly begin to show.

10.   Using the large end of a stylus, continue burnishing surface of paper. Work into recessed areas with firm pressure. Paper in recessed areas will begin to feel "soft" because the fibers are being stretched apart.

11.   Change to the small end of the stylus and continue working paper.

Caution: Make certain not to puncture paper.

12.   Carefully lift tape at edge and check design. If more definition is desired, retape and repeat Step 11.

13.   When design is satisfactory, remove tape and stencil.

**Making Body of Card and
Adhering Dry-embossed Card Insert**

14.   Trim paper to desired size and fold into a card.

15.   Using the craft knife and a straightedge, cut window opening in front of card. Opening should be 1/4" larger at each side than the dry-embossed card design. Trim 1/2" from each edge of dry-embossed design.

16.   Apply a narrow bead of craft glue to the front outside edges of dry-embossed design. Fit card insert into window opening inside front of card and secure.

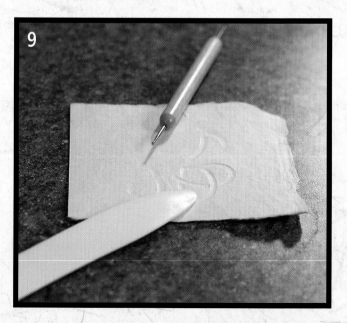

How do I add decorative elements to my handmade paper while it is still in the forming stage?

Adding texture and random patterns to handmade papers is one of the exciting aspects of this craft. Textural additions to paper are called inclusions. Items that work best must be thinner than the paper to which they are added or at least close to it in thickness. Inclusions that are too bulky can cause areas to split or tear during the couching process. Experimentation and imagination can produce spectacular results. Inclusions are added to paper in one of three methods: pulp, sheet, or vat additions.

Pulp addition: Material is added to the blender and pulsed into the pulp and water. This method works well to create a speckled or granular look. Cooked banana peel, orange peel, or colored napkins are good inclusions for this process. Materials should be added to the blender at the end of the beating cycle.

Sheet addition: This is the most creative of the methods for adding inclusions. It also gives the most control. A sheet of paper is formed either by dipping or pouring, then inclusions are placed on the wet surface before it is couched or dried. This process offers the possibilities of

What You Need To Get Started:

Supplies listed for Technique 1 on page 20

Craft glue
Craft knife
Folding bone
Ribbon
Scissors
Straightedge
Waste papers and inclusions to yield nine 8$\frac{1}{2}$" x 11" sheets:
   Light green, 8 sheets
   Olive green, 4 sheets
   Preshredded paper money, $\frac{1}{3}$ cup cut into $\frac{1}{2}$" lengths, then separated

"Here's How:"
begins on page 47.

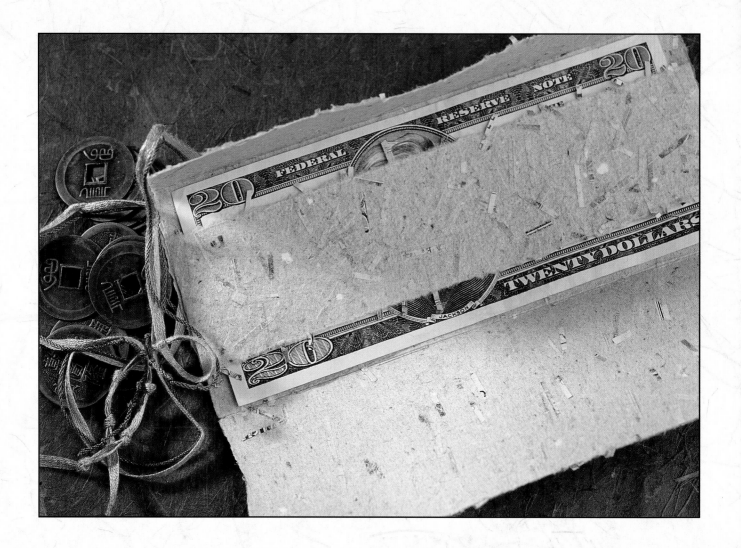

using bulkier or more heavily textured objects if more pulp is poured over the object to "seal" it to the page. Flower petals, whole flowers, leaves, and buttons can be used. Let the sheet set a few minutes before couching, then couch heavily textured sheets one at a time.

Vat addition: Materials added to the water and pulp in the vat give a random pattern and texture to the paper. They become an actual part of the paper sheet. This is an ideal way to add fibrous inclusions such as string, threads, sisal, plant fibers, or paper shred. Vat addition is the most used of the three methods.

Cash Gift Pocket

Here's How:

Making Paper Pulp and Sheets of Paper

1. Blend pulp from waste papers and cut-up preshredded paper money and pour into a plastic jug. Refer to Technique 1: Tearing and Soaking Waste Paper and Blending/Beating on pages 21 and 22.

2. Finish making the paper. Refer to Technique 1: Preparing the Work Surface and Vat on page 22, Dipping Sheets of Paper on pages 23 and 24, Couching on pages 24–27, Pressing to Remove Water on page 27, and Drying on pages 28 and 29.

Making Cash Gift Pocket

3. Using a craft knife and a straightedge, cut a 6³/₄" square from sheet of paper.

4. Measure and mark line on back side of square for the cover flap 1³/₄" from the top edge.

5. Using the end of a folding bone, gently score paper along line. Hold folding bone under paper against straightedge, then pull bone toward you, creasing paper up along straightedge. Fold paper along creased line.

6. Measure and mark line 2³/₄" down from scored and folded line.

7. Repeat Step 5. Reinforce each fold by firmly smoothing with folding bone.

8. Cut a 1¹/₂" x 7¹/₄" strip from another sheet of paper.

9. Measure and mark line for tabs ¹/₄" in on back side of each end of strip.

10. Repeat Step 7. Strip should now measure 1¹/₂" x 6³/₄" with tabs folded under.

11. Place strip, face up and tabs out, on work surface. Apply craft glue to front side of each tab. Fold tabs under, then center and attach to middle section of cover.

12. Insert cash in pocket. Fold bottom flap up, then top flap down to cover pocket. Tie with ribbon.

How do I make patterns within a sheet of handmade paper using wet embossing?

Embossing is a decorative technique used to give a raised pattern to the surface of a piece of paper. More pronounced than texturing, it adds another dimension to a flat sheet.

Wet embossing is worked into wet pulp as the sheet is made or on a sheet which has been couched and pressed to remove some of the water.

A number of different methods can be used alone or together to make an embossed image. The most basic method is to press an object into the damp sheet and then remove it. Another technique is to couch a sheet of paper onto a textured object, press paper and object under weights, and dry.

Embossed sheets are generally pressed and dried individually so the pattern is not imprinted into surrounding sheets.

**What You Need
To Get Started:**

Supplies listed for
Technique 1 on
page 20

Archival-quality
   glue stick
Archival-quality
   marker
Lace
Paintbrush,
   $1/4$" stencil
Photographs
Sizing
Waste paper to yield
ten $8^1/_2$" x 11" sheets:
   Antique white,
   24 sheets

Scrapbook Page

Here's How:

Making Paper Pulp and Sheets of Paper

1. Blend pulp from waste paper and pour into a plastic jug. Refer to Technique 1: Tearing and Soaking Waste Paper and Blending/Beating on pages 21 and 22.

2. Finish making the paper. Refer to Technique 1: Preparing the Work Surface and Vat on page 22, Dipping Sheets of Paper on pages 23 and 24, Couching on pages 24–27, and Pressing to Remove Water on page 27. Do not dry sheets of paper.

3. Separate couching cloths, leaving paper intact on each cloth.

Note: The paper must remain wet in order for it to be embossed.

Wet-embossing Sheets

4. Place damp felt on pressing board as a base. Lay strip of heavily textured lace across felt. Tuck excess lace under board so it does not move, or secure to felt with straight pins at each end.

5. Place couching cloth with damp paper face down on lace. Position 1" from edge.

6. Cover with another piece of damp felt and gently press with rolling pin.

7. Remove top piece of felt and couching cloth.

8. Using a stencil paintbrush, gently tap in an up-and-down motion, over paper and lace. This will push paper fibers further into lace, enhancing the pattern.

9. Carefully transfer embossed sheet to drying board. Smooth paper to board, avoiding embossed area. Let dry.

10. Using a glue stick, glue photographs to embossed page.

11. If journaling is to be done on the page, the sheet will need to be sized first to prevent bleeding, feathering, or soaking through. Journaling can be added as desired, using an archival-quality marker.

How do I make patterns within a sheet of handmade paper using watermarks?

There is a magical quality to watermark designs. They are generally so subtle they do not show until the sheet of paper is held to a light source. A watermark is a translucent area which is the result of sewing a wire design onto the mold screen with very fine wire or thread. The raised design of the wire causes the layer of pulp formed on it to be thinner than that on the rest of the screen. More light shines through thinner paper, thus outlining the design. The design becomes a part of the paper itself, hence the theory that early watermarks were used as a type of trademark or identification.

Lamp Shade

Here's How:

Making the Watermark Design and Sewing it to the Mold

1. Using wire cutters, cut the wire into several lengths measuring 1$\frac{1}{2}$" to 2$\frac{1}{2}$".

2. Using needle-nosed pliers, bend the lengths of wire into flat spiral shapes.

What You Need To Get Started:

Supplies listed for Technique 1 on page 20

Card stock, heavy-weight
Craft glue
Craft wire, 20-gauge
Hot-glue gun and glue sticks
Lamp shade with wire form
Needle-nosed pliers
Nylon thread, fine
Opaque Mylar®
Paintbrush, #2 watercolor
Pencil
Scissors
Waste papers to yield twelve 8$\frac{1}{2}$" x 11" sheets:
    Black, 2 sheets
    Black tissue, 1 sheet
    Silver tissue, 1 sheet
    White, 4 sheets
Wire cutters

3. Using a needle and nylon thread, stitch the wire spirals securely to the screen, attaching only at key points.

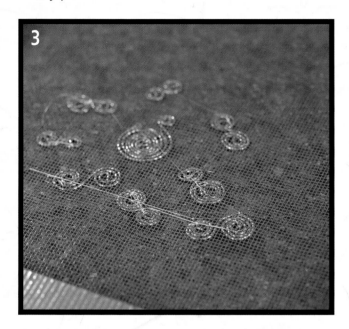

Making Paper Pulp and Sheets of Paper

4. Blend thin pulp from waste papers and pour into a plastic jug. Refer to Technique 1: Tearing and Soaking Waste Paper and Blending/Beating on pages 21 and 22.

Note: Thick pulp will not show the water-marks—it will show instead as an embossing.

5. Finish making the paper. Refer to Technique 1: Preparing the Work Surface and Vat on page 22, Dipping Sheets of Paper on pages 23 and 24, Couching on pages 24–27, Pressing to Remove Water on page 27, and Drying on pages 28 and 29.

Making Lamp Shade

6. Using the wire form of the lamp shade as a pattern, trace one section of the lamp shade onto the card stock for a template.

7. Add 1/4" margin around the outside edges of the template. Mark and cut out.

8. Arrange template on watermark paper so design will appear as desired. Mark and cut out four pieces.

9. Cut 1/4" margin from around template.

10. Placing the template onto Mylar, mark and cut out four pieces.

11. Using a watercolor paintbrush, apply craft glue to one side edge of the watermark paper.

Note: Use a damp washcloth to wipe excess glue from your fingers so the paper remains clean and untorn.

12. Overlapping 1/4", join with like edge on another piece of watermark paper. This will make two sides of the lamp shade.

13. Check the fit against the lamp shade wire form.

14. Repeat Steps 11 and 12 until pieces are joined into a flat strip and paper fits the base as desired.

15. Using a hot-glue gun, carefully adhere Mylar pieces to the wire form.

16. Fit the watermark paper strip around the Mylar "shade."

17. Using the watercolor paintbrush and craft glue, join the last open side.

18. Tuck excess paper at top and bottom of shade over the wire and glue to the Mylar.

11
technique

What You Need To Get Started:

Supplies listed for
Technique 1 on
page 20

Bamboo rods
Broom handle
Cellulose powder
Craft glue
Paintbrushes:
   $1/2$" flat
   $1^1/_2$" china bristle
   #6 liner
   #2 watercolor
#4 Pencil
Pint jar
Round template
Sheet of yellow
   handmade paper
   for the moon
Stencil
Waste papers to yield
twelve 5" x 7" sheets:
   Abaca, five 1"-squares
   Tan, 5 sheets
   White, 5 sheets
Water-based paints:
   dark brown,
   light brown,
   green, pink
Waxed paper

How do I make a large sheet of handmade paper with only a small mold?

A large piece of paper can be easily made with a small mold. This technique involves working with newly formed wet sheets, a large support board or surface, and felts. The deckle is not used in this process because the feathered edges are used to help join and bond the smaller sheets together. Large papers with unique shapes, widths, and lengths can be created with this simple process.

Scrolls had a long history in the Middle East before their appearance in China. In the third century B.C., Chinese scribes discovered they could write on long strips of silk with brushes made from animal hair. Prior to that time, their writing instruments were rigid and would have torn soft materials such as silk and paper. Today the ancient scroll form is used throughout the world for both religious and decorative purposes.

Hanging Scroll

Here's How:

**Making Paper Pulp
and Sheets of Paper**

1. Blend pulp from waste papers and pour into a plastic jug. Refer to Technique 1: Tearing and Soaking Waste Paper and Blending/Beating on pages 21 and 22.

2. Finish making the paper. Refer to Technique 1: Preparing the Work Surface and Vat on page 22, Dipping Sheets of Paper on pages 23 and 24 (the deckle will not be used), Couching on pages 24–27, and Pressing to Remove Water on page 27.

Note: Sheets of paper must remain wet in order to bond.

3. Transfer first sheet to top corner of drying board.

4. Carefully lift couching cloth.

Note: It may be necessary to peel bottom edge of wet paper from couching cloth as it is lifted. Smooth edge of paper to board. Continue lifting cloth. Once the bottom edge is sealed, the entire cloth should easily come away from the rest of the paper.

Overlapping Sheets
to Lengthen

5. Repeat Steps 3 and 4 with another sheet, overlapping each sheet approximately 1".

6. Repeat Step 5 until sheet is the desired length. Set aside and dry two extra sheets for hanging tabs.

Note: The project shown measures 8" x 33".

7. Using a rolling pin, press water from long sheet. Make certain not to drag felt over paper surface. Always lift it before moving.

8. Let dry and remove from board. Refer to Technique 1: Drying on pages 28 and 29.

Embellishing Scroll

9. Using a round template and a liner paintbrush loaded with water, mark a circle on the yellow sheet of handmade paper for the moon. Gently tear moon from paper at the "water line." Refer to Technique 6: Steps 3 and 4 on page 41.

10. In a pint jar, combine 1/2 teaspoon cellulose powder and 1 cup cool water. Stir well and let set for 20 minutes. Stir again, adding water until liquid is a syrup-like consistency.

11. Tear strips of waxed paper and slip them under edges of long sheet of paper.

12. Using a china bristle paintbrush, apply a coat of cellulose mixture over surface of paper. Let dry.

13. Repeat Step 12 on reverse side.

14. Apply cellulose mixture to one side of moon and attach to surface of scroll.

15. Using a stencil as a guide, draw in the design with a pencil.

16. With water-based paints and a watercolor paintbrush, fill in areas of the design as desired.

Finishing Scroll for Hanging

17. Tear extra sheets into eight strips, approximately 3/4 to 1" wide.

18. Using a 1/2" flat paintbrush, apply a coat of craft glue to back side of paper strip and adhere to another strip, creating four double-layer tabs.

19. Before glue has completely set, wrap tabs around broom handle to shape. Remove and let dry.

20. Glue two tabs each at top and at bottom of scroll.

21. Slip tabs over bamboo rods at top and at bottom of scroll.

**What You Need
To Get Started:**

Supplies listed for
Technique 1 on
page 20

Acrylic paints:
   pink, red, rose
Cellulose powder
Copper leafing
Craft glue
Craft knife
Découpage medium,
   matte finish
Gesso
Measuring spoons
Paintbrushes:
   1/2" flat
   Stiff-bristle
Paper towels
Plastic cup
Resin molds:
   Various faces
   and hands
Sandpaper,
   fine-grit
Table fork
Talcum powder
Waste paper to yield
one cup pulp:
   White tissue,
     5 sheets
Wooden box with lid,
   round

*"Here's How:"
begins on page 58.*

How do I make paper pulp to use as papier-maché for three-dimensional projects?

Papier-maché has an appeal and allure that is difficult to resist. Inexpensive, easy to work with, and readily available—it is an ideal medium for children or adults. Its creative possibilities are unlimited.

Three-dimensional Box

Here's How:

**Making Paper Pulp
to Use as Papier-maché**

1. Blend pulp from waste paper and pour into a plastic jug. Refer to Technique 1: Tearing and Soaking Waste Paper and Blending/Beating on pages 21 and 22.

2. Using a strainer, strain one cup processed paper pulp. Shake strainer to remove as much water as possible, but do not squeeze. Set paper pulp on a paper towel to drain. Repeat until all pulp has been strained.

3. Place strained paper pulp into a bowl. Add one teaspoon craft glue to each cup of pulp.

4. Using a table fork, mix well, then lightly flatten. Sprinkle $3/4$ teaspoon cellulose powder over the pulp. Mix well, then set aside for 15 to 20 minutes to allow the glue and cellulose powder to be absorbed into the pulp.

Note: Pulp mass will become "slick" to the touch.

5. Knead like bread dough for one minute.

6. Pinch off walnut-sized pieces of pulp. Roll pieces into balls and place balls on a dish towel.

Note: The dish towel will help pull excess moisture from the pulp and compress it.

7. Fold the corner of the towel over pulp and press to remove as much moisture as possible. Work back into balls.

Preparing the Molds
and Molding the Faces and Hands

8. Sprinkle insides of face and hand molds with talcum powder, making certain to pay special attention to the nose areas.

9. Pinch one ball of pulp into a point. Push the point into the nose area of one of the molds first, then work into surrounding areas. Push a piece of folded paper towel into the back of the mold to blot off additional moisture.

10. Turn the mold over and tap firmly to loosen the pulp. Using your fingers, gently work the molded pulp from the mold. Let dry.

11. Repeat Steps 8–10 for each molded face desired.

Note: If using a resin mold, the mold must be scrubbed with a small brush and thoroughly dried before each casting.

12. Mold the number of hands desired.

Embellishing Box and Lid

13. Using sandpaper, sand the box and lid to smooth the edges and remove imperfections. Wipe with a damp paper towel to remove excess dust.

14. Using a craft knife, trim the excess paper from the edges of the dried castings.

15. Using a 1/2" flat paintbrush, apply a light coat of découpage medium to seal the surface of molded faces and hands. Let dry.

16. Apply two coats of gesso to the surface of molded faces and hands. Let dry after each coat.

17. Apply a light coat of découpage medium to the bottom of molded faces and hands and place on lid as desired. Press to adhere. Let dry.

18. In a plastic cup, mix one tablespoon découpage medium with two teaspoons water. Stir with the handle of the paintbrush.

19. Apply the diluted découpage mixture over the molded faces and hands, making certain it is worked into all recessed areas. Let dry and repeat.

20. Randomly paint the box and lid with acrylic paints. Let dry.

Applying Copper Leaf

21. Using the 1/2" flat paintbrush, randomly apply full-strength découpage medium to small areas of painted faces and hands.

22. Brush small pieces of copper leaf onto the wet découpage medium.

23. Using a stiff-bristle paintbrush, push the copper leaf into the crevices. Lightly tap the surface of the copper leaf to achieve a "crackled" look.

24. Using the 1/2" flat paintbrush, seal the lid with a coat of découpage medium.

How do I use découpage medium to make creative and unusual projects?

Découpage is a technique which traditionally uses cut papers and printed images to cover a surface. It is then covered with a varnish-type medium to seal and protect the object.

Using torn handmade papers is a step away from the traditional process. By tearing the paper and creating a feathered edge, surfaces can be smoothly blended, giving a natural appearance and feel to the project. The great advantage comes in having to apply only two or three coats of varnish instead of the many required to fill in a cut paper edge.

Découpaged Vase

Here's How:

Making Paper Pulp and Sheets of Paper

1. Blend pulp from waste papers and pour into a plastic jug. Refer to Technique 1: Tearing and Soaking Waste Paper and Blending/Beating on pages 21 and 22.

2. Blend additional pulp from colored napkins and three cups water and pour into another plastic jug.

Note: Pulse-blend until desired consistency of color (specks) is reached. This will happen quickly so check often.

3. Combine both batches of pulp in the vat and finish making the paper. Refer to Technique 1: Preparing the Work Surface and Vat on page 22, Dipping Sheets of Paper on pages 23 and 24, Couching on pages 24–27, Pressing to Remove Water on page 27, and Drying on pages 28 and 29.

What You Need To Get Started:

Supplies listed for Technique 1 on page 20

Acrylic paints:
  Gold
  Red brown
Découpage medium, matte finish
Gesso
Hot-glue gun and glue sticks
Paintbrushes:
  1/2" flat
  #6 liner
  Stiff-bristle
Paper towels
Plastic cup
Rubbing alcohol
Sandpaper, fine-grit
Sponge
Straightedge
Vase-like container
Waste papers to yield ten 8 1/2" x 11" sheets:
  Base pulp:
    Buff, 6 sheets
    Dark green, 1 sheet
    Sage green, 3 sheets
  Dinner-sized napkins, tan (3)
Water-based marker, fine-tipped

Preparing and Painting Vase

4. Clean the surface of the vase with a paper towel and rubbing alcohol.

5. Using a marker, draw a linear-type design on the vase.

6. Using a hot-glue gun, apply hot glue over the top of the design. Let dry.

7. Wipe over the design to remove any excess marker lines which may not have been covered with hot glue.

8. Using a 1/2" flat paintbrush, apply a generous coat of découpage medium to seal the surface of the vase. Let dry.

9. Apply a coat of gesso to the surface of the vase. Let dry.

10. Sponge over the vase with gold acrylic paint. Let dry.

Texturing Vase

11. Tear handmade paper into strips approximately 1" x 4". Refer to Technique 6: Steps 3 and 4 on page 41.

12. In a plastic cup, mix two parts découpage medium with one part water. Stir with the handle of the paintbrush.

13. Apply the diluted découpage mixture over an area at the back of the vase. In a vertical position, adhere a strip of torn paper to the vase, gently brushing from the center to the outside edges.

14. Continue adhering strips, covering the raised design last.

15. Using a stiff-bristle paintbrush, gently push the paper into the raised design. Be careful not to tear damp paper. Blend the edges. Let dry.

Note: More découpage medium may be applied as necessary to keep the paper workable.

16. Using sandpaper, lightly sand the surface of the vase. Remove excess dust with a damp paper towel.

17. Using the 1/2" flat paintbrush, apply two coats of full-strength découpage medium to the surface of the vase. Let dry after each coat.

18. Using your index finger, apply gold acrylic paint to the raised areas of the design. Let dry.

19. Working small areas at a time, dry-brush red brown paint around edges of raised design. Remove excess paint with a damp paper towel.

20. Using the 1/2" flat paintbrush, apply a generous coat of découpage medium to the surface of the vase. Let dry.

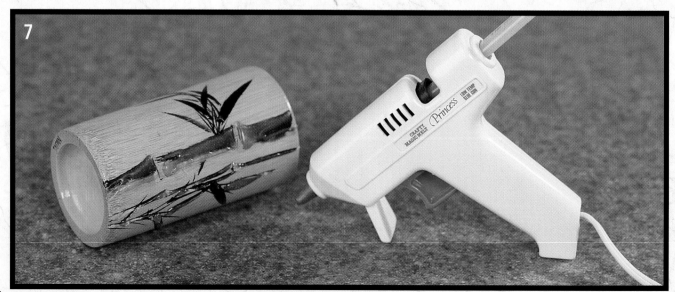

How do I make one sheet of handmade paper a different color on each side by laminating?

Creating a sheet of paper with a duplex surface—one that is a different color on each side—is simple and fun. The most difficult part of the process is deciding which colors to use. Colors may be high contrast, such as red and black, or soft and subtle pastels. Each side of the sheet will maintain its color integrity because of the way it is formed. Two vats will be needed for this project.

**What You Need
To Get Started:**

Supplies listed for
Technique 1 on
page 20

Cord
Craft knife
Decoration, small
Folding bone
Pencil
Straightedge
Waste papers to yield
nine 8$\frac{1}{2}$" x 11"
laminated sheets:
   Deep purple tissue,
    3 sheets
   Hot pink tissue,
    3 sheets

Origami Folder

Here's How:

**Making Paper Pulp
and Sheets of Paper**

1. Blend very thin pulp from each color of waste papers and pour into separate plastic jugs. Refer to Technique 1: Tearing and Soaking Waste Paper and Blending/Beating on pages 21 and 22.

2. Continue making the paper. Refer to Technique 1: Preparing the Work Surface and Vat on page 22, Dipping Sheets of Paper on pages 23 and 24, and Couch-ing on pages 24–27. Couch sheet of first colored pulp onto couching mound. Repeat couching sheet from second colored pulp directly on top of first sheet. Take care to match edges of sheets as closely as possible. Cover with couching cloth.

3. Finish making the paper. Refer to Technique 1: Pressing to Remove Water on page 27, and Drying on pages 28 and 29.

Note: The sheets laminate or "stick" to each other as they dry.

Design Tip: *Laminated papers can vary widely in thickness and in texture. This project works best with a thin, smooth-surfaced paper. It is advisable to practice the folds with typing or copier paper first.*

Folding Origami Folder

4. Using a straightedge and a craft knife, trim one laminated sheet of paper to 8" x 10½".

5. Using the straightedge and a pencil, mark the measurements as shown on Diagram A below onto the paper.

6. Using the end of a folding bone, score folder.

7. With darker side of paper facing up, crease as shown in Diagram B below.

8. Following crease lines, fold left edge in and right edge up placing it over left edge as shown in Diagram C below.

9. Measuring 9" from top point, fold bottom part under as shown in Diagram D below.

10. Secure with cord and a small decoration approximately 1" up from the bottom fold.

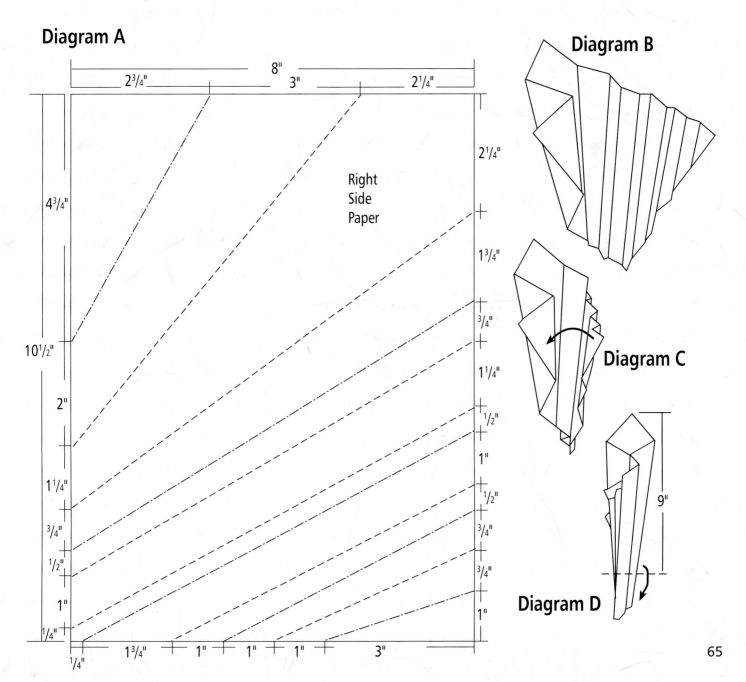

Diagram A

Diagram B

Diagram C

Diagram D

Right Side Paper

Section 3: *beyond the basics*

68

How do I add sizing to my handmade papers and burnish them?

Burnishing produces a lovely, glowing sheen on the paper just as it does on other suitable materials. A smooth, hard object is rubbed over the paper's surface to press down and bond the fibers into a tight mat. In the very early days of papermaking, burnishing was the final step toward producing a smooth, acceptable writing surface.

What You Need To Get Started:

Supplies listed for Technique 1 on page 20

Craft glue
Folding bone
Gelatin, unflavored
Measuring cup with pouring spout
Measuring spoons
Newspaper
Plastic container, approximately 10" x 13"
Scrap paper
Soap bars
Spray starch, heavy-duty
Waste paper to yield four 8½" x 11" sheets: Light tan, 12 sheets

Soap Wraps

Here's How:

Making Paper Pulp and Sheets of Paper

1. Blend pulp from waste paper and pour into a plastic jug. Refer to Technique 1: Tearing and Soaking Waste Paper and Blending/Beating on pages 21 and 22.

2. Finish making the paper. Refer to Technique 1: Preparing the Work Surface and Vat on page 22, Dipping Sheets of Paper on pages 23 and 24, Couching on pages 24–27, Pressing to Remove Water on page 27, and Drying on pages 28 and 29.

Adding Sizing

3. Size the sheets with gelatin. Refer to Technique 4: Adding Sizing on pages 36 and 37.

4. Size the sheets with spray starch. Refer to Technique 5: Adding Sizing on page 39.

Burnishing Sheets

5. Place pad of newspaper on a hard, flat surface. Cover with a sheet of clean scrap paper.

6. Using the end of a folding bone and beginning at the outside edge of a sheet of paper, press firmly and rub in an overlapping circular motion until the entire surface of one side of the sheet is completely burnished. Repeat for each sheet.

Making Soap Wraps

7. Tear handmade paper into strips and wrap around soap bars. Refer to Technique 6: Steps 3 and 4 on page 41.

8. Overlap the edges at the back of each bar and adhere with craft glue.

2

project

What You Need To Get Started:

Supplies listed for Technique 1 on page 20

Assorted decorative
   elements
Baking soda
Bananas (5)
Craft scissors
Découpage medium,
   matte finish
Dishwashing detergent
Industrial-strength glue
Measuring cup with
   pouring spout
Measuring spoons
Metallic copper thread
Paintbrushes:
   1/2" flat
   #6 liner
Plastic cup
Rocks
Saucepan
Scissors
Straightedge
Table knife
Toothbrush
Waste papers to yield
twelve 8 1/2" x 11" sheets:
   Light tan, 4 sheets
   Purple, 4 sheets
   White, 8 sheets
Water-based
   satin varnish

How do I add color to my handmade papers using natural ingredients?

Natural ingredients can be used effectively in re-cycled papers to enhance their handmade appearance. There is always an element of surprise involved when using such things as beet juice—a deep red fades to a hint of pink at the edges of the paper; orange rinds add a lovely, subtle scent to the paper; and banana peels—the paper can be "chunky" or have a heavy, sandy-looking speckle with mauve undertones.

Paperweights

Here's How:

Preparing Banana Peels

1. Remove the peels from five bananas. Remove and discard both ends.

2. Lay peels flat with membrane sides up. Using a table knife, gently scrape each peel to remove the pulp. Discard pulp. Continue scraping until the banana peel is completely exposed.

3. In a saucepan, add banana peels to two cups boiling water. Add two teaspoons baking soda and simmer for 10 minutes.

4. Using a strainer, strain peels and rinse.

**Making Paper Pulp
and Sheets of Paper**

5. Blend pulp from waste papers and pour into a plastic jug. Refer to Technique 1: Tearing and Soaking Waste Paper and Blending/Beating on pages 21 and 22.

6. Blend banana peels with two cups water and add to the paper pulp as desired.

Note: For larger "chunks" of banana fiber, pulse three or four times; for a finer texture, blend at high speed for five to seven seconds.

7. Finish making the paper. Refer to Technique 1: Preparing the Work Surface and Vat on page 22, Dipping Sheets of Paper on pages 23 and 24, Couching on pages 24–27, Pressing to Remove Water on page 27, and Drying on pages 28 and 29.

Preparing the Rocks

8. Using a toothbrush and mild dishwashing detergent, scrub the rocks. Rinse well, and let dry.

9. Using a ½" flat paintbrush, apply two light coats of varnish to the rocks. Let dry between coats.

**Making Paper Strips and
Adhering Them to Rocks**

10. Tear handmade paper into strips approximately 1" x 8" . Refer to Technique 6: Steps 3 and 4 on page 41.

11. In a plastic cup, mix one part découpage medium with one part water. Stir with the handle of the paintbrush.

12. Using the ½" flat paintbrush, apply the diluted découpage mixture to the back side of each paper strip. Arrange paper strip(s) on each rock and brush to adhere in place, applying more découpage mixture as necessary. Let dry.

Note: If the découpage mixture bleeds onto the glossy surface of the varnished rocks it will leave a dull "halo." Instead of trying to remove the halo, let it dry. Then, using the liner paintbrush and a drop of varnish, touch-up the dull area by brushing from the surface of the rock just to the edge of the paper.

Embellishing Paperweights

13. Wind the metallic copper thread in random patterns around each rock. Cut and tie the ends of the thread together on the bottom of each paperweight.

14. To make a smooth surface on the bottom of each paperweight, cover the thread with small pieces of handmade paper and apply the découpage mixture over the top to adhere the paper to the thread and the paperweights. Let dry.

15. Glue assorted decorative elements such as beads and tassels onto each paperweight.

How do I add color to my handmade papers using paint and decorative techniques?

A characteristic of handmade paper is its acceptance of so many types of surface decoration. Sheets can have additions made to them in the couching process, before they are pressed and dried, or the more traditional approach can be used—decorating or painting after the sheet has dried.

As choices are made about additions to a paper surface, it is advisable to take into consideration the weight of and sizing in the background sheet. Damp or wet additions, such as glue or paint, can wrinkle and weaken a thin sheet beyond repair. Heavier papers may require more sizing so glue will adhere to the surface rather than being absorbed into the sheet.

Wrapped Vase

Here's How:

Making Paper Pulp and Sheets of Paper

1. Blend pulp from waste paper and pour into a plastic jug. Refer to Technique 1: Tearing and Soaking Waste Paper and Blending/Beating on pages 21 and 22.

2. Reserve approximately two tablespoons paper pulp to be used later.

3. Finish making the paper. Refer to Technique 1: Preparing the Work Surface and Vat on page 22, Dipping Sheets of Paper on pages 23 and 24, Couching on pages 24–27, Pressing to Remove Water on page 27, and Drying on pages 28 and 29.

3
project

What You Need To Get Started:

Supplies listed for Technique 1 on page 20

Acrylic paints
Confetti
Craft glue
Craft wire,
   28-gauge
Glitter
Liquid paper adhesive
Measuring spoons
Paintbrushes:
   $1/2$" flat
   #2 watercolor
Paper towels
Plastic cup
Plastic wrap
Spray starch,
   heavy-duty
Tape, low-tack
Vase
Waste paper to yield six $8^1/_2$" x 11" sheets:
   White, 12 sheets
Wire cutters

Adding Sizing

4.   Size the sheets with spray starch. Refer to Technique 5: Adding Sizing on page 39.

Making Decorative Pulp

5.   In a plastic cup, mix two tablespoons reserved, drained paper pulp, a sprinkle of glitter, confetti, and four to five drops of craft glue. Stir well until mixture feels slightly sticky.

6.   Place one to two teaspoons pulp mixture between two sheets of plastic wrap and flatten with the palm of your hand.

7.   Using a rolling pin, flatten to a paper-like consistency. If the mixture squeezes out at the edges of the plastic wrap, return excess to the plastic cup and wipe up with a damp paper towel.

8.   Remove the top sheet of plastic wrap and set flattened pulp aside to dry. This may take several hours.

9.   When decorative pulp is dry, break or tear it into random shapes and sizes.

Embellishing Sheet of Paper and Wrapping Vase

10.   Using a 1/2" flat paintbrush, apply liquid paper adhesive to the dried decorative pulp shapes and adhere them to one of the sized sheets of paper, leaving some open space.

11.   Using a watercolor paintbrush and acrylic paints, fill in the open spaces of the design. Let dry. Repeat until desired color is achieved.

12.   Wrap the decorated sheet around vase and secure with low-tack tape.

Note: Sticky tapes can easily tear or abrade softer papers when removed.

13.   Wrap wire around the vase and twist it tightly to secure, taking care not to break it.

14.   Using wire cutters, cut off excess, leaving a 3/4" tail. Tuck the tail under the wires. Remove the tape.

How do I add decorative elements to my handmade papers using natural inclusions?

Finding natural inclusions to add to sheets of paper is as easy as stepping to the refrigerator or the spice cabinet. For this unusual project, small peppers were sliced thin and dried between the pages of an old phone book. Dried, crushed skins and seeds were also added.

What You Need To Get Started:

Supplies listed for Technique 1 on page 20

Acrylic paints:
    Gold
    Red
Aluminum foil
Bell pepper, fresh
Craft glue
Découpage medium
Paintbrush,
    ¹/₂" flat
Plastic cup
Thin twigs, the number
    to match
    the crevices
    on the bell pepper
Tissue paper:
    White, 2 sheets
Waste paper and
inclusions to yield
nine 8¹/₂" x 11" sheets:
    Dried pepper skins
        and seeds, ³/₄ cup
    White, 12 sheets

Pepper Pot

Here's How:

Making Paper Pulp and Sheets of Paper

1. Blend pulp from waste paper and dried pepper skins and seeds and pour into a plastic jug. Refer to Technique 1: Tearing and Soaking Waste Paper and Blending/Beating on pages 21 and 22.

2. Finish making the paper. Refer to Technique 1: Preparing the Work Surface and Vat on page 22, Dipping Sheets of Paper on pages 23 and 24, Couching on pages 24–27, Pressing to Remove Water on page 27, and Drying on pages 28 and 29.

Creating Pepper Pot

3. Tear one sheet of tissue paper into pieces approximately 1" square.

4. Wrap bell pepper in aluminum foil, leaving top edge of pepper uncovered. Using your finger, gently smooth out ridges.

5. In a plastic cup, mix one tablespoon découpage medium with two teaspoons water. Stir with the handle of the paintbrush.

6. Using a 1/2" flat paintbrush, apply the diluted découpage mixture over the form. Randomly place the pieces of tissue paper over the découpage mixture and gently brush over to adhere. Continue until all aluminum foil is covered with one layer of tissue paper. Let dry. Repeat this process two more times. Let dry between each layer.

Note: When dry, a firm, but flexible pot is formed.

7. Remove the pepper from the tissue paper pot. Remove excess aluminum foil where necessary.

8. Tear handmade paper into randomly sized pieces. Refer to Technique 6: Steps 3 and 4 on page 41.

9. Apply diluted découpage mixture over the top layer of tissue paper. Randomly place the pieces of torn handmade paper over the découpage mixture and gently brush over to adhere. Continue until the entire pot is covered with one layer of handmade paper. Let dry. Repeat this process. Let dry.

10. Tear remaining sheet of tissue paper into strips approximately 1" x 5".

11. Using the 1/2" flat paintbrush, apply the diluted découpage mixture to one side of one strip of tissue.

12. Wrap one twig with tissue. Smooth out rough areas and edges. Let dry.

13. Repeat Steps 11 and 12 for each twig.

Painting Pepper Pot

14. Using the 1/2" flat paintbrush, apply two generous coats of red paint to the interior of the pot. Let dry between each coat.

15. Apply one light coat of gold paint to the interior of the pot. Let dry.

16. Apply two generous coats of gold paint to the twigs. Let dry between each coat.

17. Press one twig into each crevice on the pot and adhere with craft glue. Make certain to adjust twigs as necessary so pot will stand.

How do I add decorative elements to my handmade papers using synthetic inclusions?

Synthetic inclusions include any "man-made" decorative additions added to paper in its forming stages. Synthetic inclusions require some considerations which natural additions do not. They work best when added to the pulp water in the vat. This process covers the inclusions with a thin layer of fibers, holding the additions in place and actually making them a part of the sheet.

5
project

What You Need To Get Started:

Supplies listed for Technique 1 on page 20

Acrylic paint, black
Card stock, black
Craft knife
Découpage medium
Musical note
    paper punch, small
Paintbrush, $1/2$" flat
Pencil
Picture frame
Scissors
Straightedge
Waste paper and inclusions to yield seven $8^1/2$" x 11" sheets:
    White, 12 sheets
    Black embroidery
        floss, 1 yard
        cut into $1/2$" to
        1" lengths,
        then separated
Water-based
    gloss varnish

Picture Frame

Here's How:

Making Paper Pulp and Sheets of Paper

1. Blend pulp from waste paper and embroidery floss and pour into a plastic jug. Refer to Technique 1: Tearing and Soaking Waste Paper and Blending/ Beating on pages 21 and 22.

2. Using a small musical note paper punch, punch $1/4$ cup music note shapes from black card stock and add to the pulp as a decorative element.

3. Finish making the paper. Refer to Technique 1: Preparing the Work Surface and Vat on page 22, Dipping Sheets of Paper on pages 23 and 24, Couching on pages 24–27, Pressing to Remove Water on page 27, and Drying on pages 28 and 29.

Preparing Picture Frame

4. Using a $1/2$" flat paintbrush, apply two coats of black acrylic paint to the frame. Let dry after each coat.

5. Apply two coats varnish to the frame. Let dry after each coat.

6. Using a straightedge and a pencil, measure and mark the sheet of paper to fit the front surface of the frame minus 1" on each side. Measure and mark opening.

7. Using a craft knife, cut out the opening.

8. Using the $1/2$" flat paintbrush, apply a light coat of découpage medium to back side of paper and adhere to surface of frame.

How do I make shaped handmade papers—without cutting or tearing them— using a shaped mold or deckle?

A piece of paper can be made in any shape as long as a mold can be made to create it on. The simplest shapes to work with, beyond rectangles and squares, are ovals and circles. An embroidery hoop covered with two layers of fine nylon tulle is an ideal, temporary mold.

The most difficult part of making an envelope deckle is deciding what shape to make. They are so easy and inexpensive to do, it is worth trying several styles. To make envelopes, first measure the width of the paper to be used as stationery and add $1/2"$. This is how wide the envelope must be so the paper will fit into it correctly.

Stationery

Here's How:

Making Paper Pulp and Sheets of Paper

1. Blend pulp from waste papers and pour into a plastic jug. Refer to Technique 1: Tearing and Soaking Waste Paper, Blending/ Beating on pages 21 and 22, and Preparing the Work Surface and Vat on page 22.

2. Tear comics into $1/2"$ squares. Pack and fill three cups. Add one cup torn comics to the vat. Mix well.

Note: Newspaper comics produce a lighter and brighter paper than comic books.

3. After every five sheets, replenish the vat by adding an additional cup of torn comics.

What You Need To Get Started:

Supplies listed for Technique 1 on page 20

Craft glue
Craft knife
Duct tape
Foam sheeting
Marking pen
Newspaper comics, colored
Nylon tulle
Ready-made envelope
Spray starch
Straightedge
Waste papers to yield fifteen 6" x 10" oval sheets:
    Pale pink, 3 sheets
    White, 9 sheets
Waste papers to yield twelve 4" x $8^3/4"$ envelopes:
    Pale pink, 3 sheets
    White, 9 sheets
Wooden embroidery hoop, 6" x 10" oval

4. Finish making the paper using an oval mold. Refer to Technique 1: Dipping Sheets of Paper on pages 23 and 24, Couching on pages 24–27, Pressing to Remove Water on page 27, and Drying on pages 28 and 29.

Adding Sizing

5. Size the sheets with spray starch. Refer to Technique 5: Adding Sizing on page 39.

Envelope

Here's How:

Making Envelope Deckle

1. Choose a ready-made envelope to fit stationery size. Carefully pull apart glued flaps and lay flat. The dismantled envelope will become the pattern.

Note: Envelope shape must fit mold and deckle size with a minimum 1/4" border all around. If a suitable size cannot be found, enlarge or reduce it on a copy machine.

2. Place pattern on foam sheeting and trace around the shape with a marking pen. Remove pattern.

3. Using a straightedge and a craft knife, cut the envelope shape from the foam. Do not discard the "foam envelope." It makes a good pattern for later use and when taped back into the "foam deckle," it helps the foam retain its shape during storage.

4. Secure the foam deckle to the mold at all four outside edges with duct tape. Place deckle frame over the foam deckle to further secure at edges.

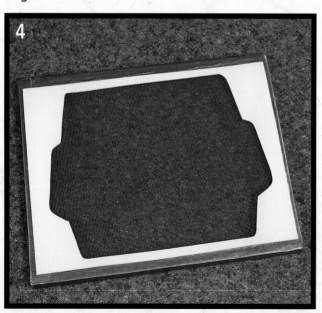

Making Paper Pulp and Envelopes

5. Blend pulp from waste papers and pour into a plastic jug. Refer to Technique 1: Tearing and Soaking Waste Paper, Blending/Beating on pages 21 and 22, and Preparing the Work Surface and Vat on page 22.

6. Tear comics into 1/2" squares. Pack and fill three cups. Add one cup torn comics to the vat. Mix well.

7. After every four envelopes, replenish the vat by adding an additional cup of torn comics.

8. Continue making the paper using the envelope deckle. Refer to Technique 1: Dipping Sheets of Paper on pages 23 and 24. Using your fingers, remove excess pulp at envelope edges as pulp will be sitting on top of the foam.

9. Finish making the paper. Refer to Technique 1: Couching on pages 24–27, Pressing to Remove Water on page 27, and Drying on pages 28 and 29.

Adding Sizing

10. If envelopes are to be hand-addressed, size the envelopes with spray starch. Refer to Technique 5: Adding Sizing on page 39.

Note: If labels are to be used for addressing, this step can be skipped. In fact, labels are a good idea because of the size of the inclusions in this particular paper recipe.

Folding Envelopes

11. Using the original envelope as a pattern, fold the handmade paper into envelopes and secure with craft glue.

How do I make patterns within my handmade papers using papermaker's tears?

Papermaker's tears are considered by many to be a mistake, which they can be if they occur in unplanned or unwanted places. Tears are the result of water droplets falling onto a newly formed sheet of paper usually from a deckle frame as it is being removed from the mold. When the water droplets hit the wet fibers of the paper, fibers separate and spread making a thin spot on the sheet. This accidental process also can be used to create a lacy, decorative effect in thinner papers.

What You Need To Get Started:

Supplies listed for Technique 1 on page 20

Cotton swab
Découpage medium
Glass votive holder
Gold leaf
Paintbrushes:
    1/2" flat
    #6 liner
Paper towels
Plastic cup
Rubbing alcohol
Straightedge
Waste papers to yield nine 8 1/2" x 11" sheets:
    Gold leaf, 1 sheet, 5 1/2" x 5 1/2"
    Lavender, 3 sheets
    Royal blue, 3 sheets

"Here's How:"
begins on page 82.

Votive Holder

Here's How:

**Making Paper Pulp
and Sheets of Paper**

1.   Blend pulp from waste papers and pour into a plastic jug. Refer to Technique 1: Tearing and Soaking Waste Paper and Blending/Beating on pages 21 and 22.

2.   Continue making the paper. Refer to Technique 1: Preparing the Work Surface and Vat on page 22 and Dipping Sheets of Paper on pages 23 and 24.

3.   Place the mold over the vat. Using a handful of clean tap water, splash it onto the dipped sheet. The impact of the water droplets will create the necessary holes or "papermaker's tears" in the sheet.

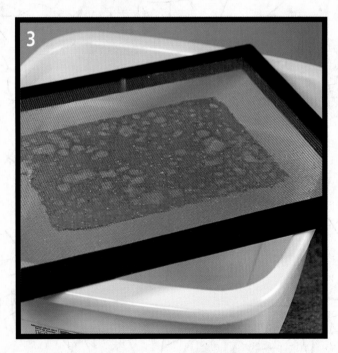

4.   Finish making the paper. Refer to Technique 1: Couching on pages 24–27, Pressing to Remove Water on page 27, and Drying on pages 28 and 29.

Applying Gold Leaf

5.   Using rubbing alcohol and a paper towel, thoroughly clean the outside of the votive holder.

6.   Tear one sheet of gold leaf into $1\frac{1}{2}$" squares. Set aside.

7.   Using a $\frac{1}{2}$" flat paintbrush, apply découpage medium to rim of votive holder covering an area approximately the same size as a small piece of gold leaf. Clean the paintbrush on a paper towel.

8.   Pick up a piece of gold leaf with the paintbrush and adhere it to the wet découpage medium. Gently brush, allowing some breaks and wrinkles in the gold leaf for added interest. Let dry.

Making Paper Strips and Adhering Them to Votive Holder

9.   Tear handmade paper into strips approximately $2\frac{1}{2}$" x 11". Refer to Technique 6: Steps 3 and 4 on page 41.

Note: One long strip may be sufficient if paper is large enough. Wrap sheet around votive holder to check size before marking and tearing.

10.   In a plastic cup, mix one part découpage medium with one part water. Stir with the handle of the paintbrush.

11.   Using the $\frac{1}{2}$" flat paintbrush, apply the diluted découpage mixture to the back side of each paper strip. Arrange paper strip(s) around votive holder and brush to adhere in place, applying more découpage mixture as necessary. Let dry.

Note: Leave at least 1" rim of gold leaf showing around votive holder. Do not overlap or layer the paper strips as the design will not show through.

12.   Using rubbing alcohol and a cotton swab, clean the découpage mixture from areas where glass is exposed.

How do I use cockled paper as papier-maché for three-dimensional projects?

Sometimes a sheet of paper dries and simply will not flatten. Cockles, wrinkles, or puckers remain even after pressing. Rather than throwing the sheet away, tear it into small pieces and use them in papier-maché projects. Handmade papers are ideal for papier-maché because of their softness and the way they accept paste and glues.

8
project

What You Need To Get Started:

Supplies listed for Technique 1 on page 20

Aluminum foil
Assorted decorative
    elements
Craft glue
Craft knife
Découpage medium
Fabric
Glitter, ultrafine
Paintbrushes:
    $1/2$" flat
    #6 liner
Petroleum jelly
Plastic cups
Polyester filling
Scissors
Shoe form
Straightedge
Tissue paper:
    White, 1 sheet
Waste papers to yield
nine $8^{1}/_{2}$" x 11" sheets:
    Red tissue, 1 sheet
    Scarlet red, 8 sheets

Ruby Slipper Pincushion

Here's How:

Making Paper Pulp and Sheets of Paper

1. Blend pulp from waste papers and pour into a plastic jug. Refer to Technique 1: Tearing and Soaking Waste Paper and Blending/Beating on pages 21 and 22.

2. Finish making the paper. Refer to Technique 1: Preparing the Work Surface and Vat on page 22, Dipping Sheets of Paper on pages 23 and 24, Couching on pages 24–27, Pressing to Remove Water on page 27, and Drying on pages 28 and 29.

Creating Shoe

3. Tear tissue paper into pieces approximately 1" square.

4. Rub shoe form with a small amount of petroleum jelly to help hold the foil in place and protect against possible glue leaks.

5. Wrap shoe form in aluminum foil. Using your finger, gently smooth out ridges.

6. In a plastic cup, mix one tablespoon découpage medium with two teaspoons water. Stir with the handle of the paintbrush.

Design Tip: *Glitter adds a whimsical touch to many types of projects, but it is a nuisance when added to the pulp in the blender or in the vat. Add the "glitz" with spray glitter when embellishing sheets of paper and with glitter paint for dimensional projects.*

7. Using a ½" flat paintbrush, apply the diluted découpage mixture over the form. Randomly place the pieces of tissue paper over the découpage mixture and gently brush over to adhere. Continue until the entire form is covered with one layer of tissue paper. Let dry. Repeat this process three more times. Let dry between each layer.

Note: When dry, a firm, but flexible shoe is formed.

8. Using a craft knife, make a slit down the back of the shoe, then remove the form from the tissue paper shoe. Remove excess aluminum foil where necessary.

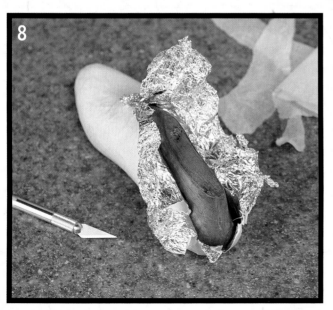

9. Push "slit" area in back of shoe together and patch with tissue and the diluted découpage mixture. Let dry.

10. Tear handmade paper into 1" to 1½" squares. Refer to Technique 6: Steps 3 and 4 on page 41.

11. Apply diluted découpage mixture over the top layer of tissue paper. Randomly place the pieces of torn handmade paper over the découpage mixture and gently brush over to adhere. Continue until the entire shoe is covered with one layer of handmade paper. Let dry. Repeat this process. Let dry.

12. In a plastic cup, mix ½ teaspoon découpage medium with two or three drops water. Stir with the handle of the paintbrush.

13. Add a generous sprinkle of ultrafine glitter to the découpage mixture and stir well.

14. Generously apply one coat of the glitter/découpage mixture to the shoe. Let dry. Repeat this process four more times. Let dry between each coat.

Making Pincushion

15. Cut an oval from the fabric measuring the length of the shoe and twice the shoe's width.

16. Gather the fabric at the outside edge with a running stitch. Pull thread to form a bag. Tie off, leaving a 1" opening for stuffing.

17. Firmly stuff with polyester filling. Push the cushion into the shoe to make certain it fits well. If the shape and size of the bag are suitable, remove the cushion and sew the opening shut. If necessary, make adjustments.

18. Place several drops of glue inside the shoe along the open edges. Push the cushion into the shoe and apply gentle pressure to adhere.

19. Add assorted decorative elements as desired.

Art piece by Linda Gunn

Art piece by Rhonda Rainey

Art piece by Jackie Abrams   Photo by Jeff Baird

Art piece by Karen Simmons   Photo by Margot Geist

Art piece by Jackie Abrams   Photo by Jeff Baird

Section 4:
gallery

Art piece by Barbara Fletcher   Photo by Jan Bindas

Art pieces by Cleo Campos Hill

In 1996, Rainey left the public school system to pursue her art career full time. She worked as an in-house designer/craftswoman at Chapelle, Ltd., where she continued to expand her knowledge of fine crafting. Rhonda has continued her association with Chapelle, Ltd., as a free-lance designer and author. Her published books include *Heartfelt Ways to Say Goodbye, Designing with Paper and Paint,* and *Faux Finishing for the first time.* Her paintings, paper art, and craft work also appear in a number of publications.

The author currently resides in Pocatello, Idaho, where she maintains a studio, mentors fellow artists, and enjoys her children, their spouses, and her grandchildren.

Rhonda Rainey

Rhonda Rainey was born and raised in southeastern Idaho. Her parents and grandparents offered encouragement and opportunities to develop her gift in the visual arts and in music.

After graduating from Ricks College and Brigham Young University with degrees in studio art, art history, and vocal music, Rainey attended Idaho State University where she earned her Idaho teaching credential in art education. In 1976 she began her public school teaching career. She worked as an artist educator for 20 years teaching all areas of the visual arts and crafts.

Summer vacations with her husband and three children were often planned around displaying and selling her award-winning watercolors at art shows throughout the western United States.

"Carnival Mask." Papier-maché, acrylic paint, and glass beads. 9" high x 7" wide.

"The Victorian Butterfly Book." Solvent transfer burnished handmade papers.
15" high x 8" wide.

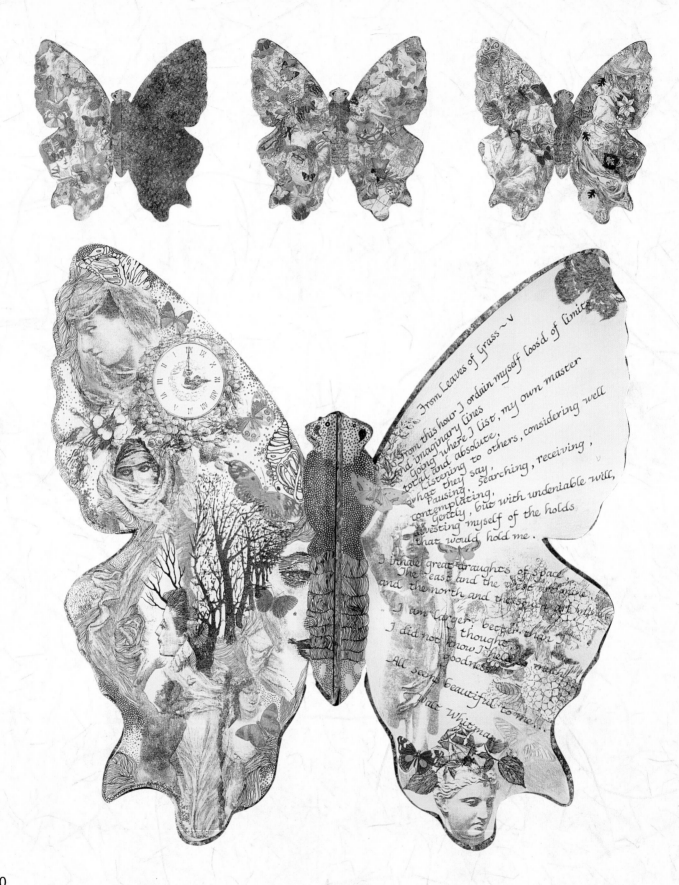

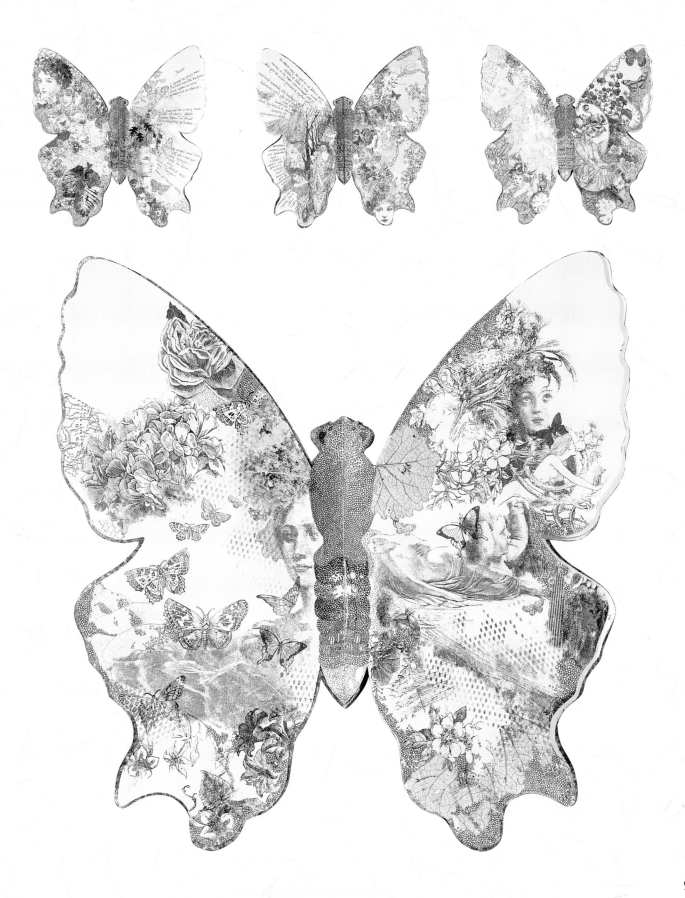

Barbara Fletcher

There is something about the tactile sensation of fiber/fabric/paper that is very satisfying and meditative for Barbara Fletcher. When she began to work with fiber over 15 years ago, she knew this was the field for her. Over the years her experimentation utilized many methods of dying, painting, batiking, and photographic techniques.

Initially, Fletcher began to make sculpture in fabric in a whimsical way. She was and still is attracted to the figure, and animals in particular. In the mid 1980s after taking a fantasy illustration course at the Rhode Island School of Design, her animals began to personify. This metamorphosis culminated in a wonderful commission of fantasy creatures for writer Stephen King. Animals, insects, fish, birds, and other creatures combined to make fantastical forms that hung suspended over an indoor pool and on the wall of the hallway leading to the pool area. She called it her Creature Evolution series. One of the major influences on her art is a sixteenth-century Dutch painter, Hieronomus Bosch. His mythical creatures are renowned for their satiric and even religious content.

Eventually, Fletcher left Maine where she had been living, moved to Boston, and began to support herself as a full-time artist. While living in Maine, she had taken a papermaking workshop at Haystack Mt. Craft School with some well-known papermakers. There, she discovered paper casting. She loved it and decided to make a switch from time-consuming fabric.

Photo by Jan Bindas

"Chameleon." 3-D paper casting. Cast in four sections, airbrushed with fabric dyes.
13" high x 19" wide x 10" deep.

Paper casting is a method of creating a clay sculpture, then making a plaster cast from the clay. With the plaster cast finished, wet paper pulp can be pressed inside. As the paper dries and shrinks, it pops out of the mold. Casting is a wonderful way of making limited production work. Fletcher was soon producing more realistic and humorous animal masks, ornaments, and three-dimensional cards. In 1995 she began to sell her cards to a company in Japan. They eventually bought thousands of them and the connection lasted for several years, until the downturn in their economy. Also at this time, she enjoyed working with art consultants, making one-of-a-kind wall pieces for hospital interiors, pediatrics wards, and even restaurants.

Most recently, Fletcher has been exploring different techniques. She is fascinated with combining light and paper texture. She has begun a series of lighted sculptures, using a technique of spray pulping. She begins with forming a wire sculpture. The wire sculpture is then sprayed with wet paper pulp blown from a spray gun under high pressure. The wire sculpture is then placed over a regular lamp shade and used with a low-watt bulb. The end result is a luminescent sculpture. Several photographs of her sculptures will be seen in a book entitled *Paper Illuminated*. She is encouraged by the response she has received in this new direction and plans on continued exploration.

Photo by Gordon Bernstein

"Monkey." Cast paper, single mold. Colored pulp and acrylic paint. 10" high x 7" wide x 5" deep.

*"Lion." Cast paper, two molds. Airbrushed with fabric dyes.
20" high x 21" wide x 10" deep.*

Photo by Gordon Bernstein

"Fish and Duck Lamps." Paper pulp blown out under high pressure
over wire sculpture. Painted with acrylics.

Colleen Barry

Colleen Barry holds a Master of Fine Arts degree from the Edinboro University of Pennsylvania. Her work is included in numerous museums, as well as corporate and private collections in the United States, Germany, Mexico, Finland, Italy, Monaco, Canada, and Japan, and has been published in several art books. She has taught at the university level and given workshops and lectures throughout the United States.

Barry creates sculptural books, shrines, and reliquaries from self-fabricated and 'found' objects. Her sculptures are narrative, storytelling vehicles containing symbols of personal and universal mythology. She thinks of them as visual journals and hopes they impart a sense of allegory, wonder, and humor.

"White Button Book." Wall piece of handmade linen/cotton papers layered and sewn in signatures to form an open book. The center pages have an embossed, painted grid. All buttons are sewed on. Measuring tapes, pins, needles, etc., are glued onto the pages.
12" high x 30" wide x 4" deep.

"Baroque Diary." Book sculpture of handmade linen/cotton papers fabricated into a closed book format with a hand-stitched and embroidered binding. Layered among the pages are assorted found objects. 6" high x 9" wide x 3" deep.

Karen Simmons

A handweaver from the early 1970s to the early 1990s, Karen Simmons has always been drawn to vessels. She began experimenting with papermaking in 1995 and found it to be an exciting and responsive medium. Her fiber background has allowed her to incorporate weaving and basketry techniques into the embellishments of her paper forms.

Always a collector of found objects, Simmons has thoroughly enjoyed the opportunity to use bones, feathers, vines, and other materials to amplify the connection between her organic shapes and their inspiration—nests, cocoons, seeds, and pods. The papers used to construct these vessels are made primarily from plant materials found around her home in the New Mexico mountains. Dry fibers are gathered in the late summer and fall. Her process is focused on collaborating with the spirit of nature at its ebb by reconstructing these fibers into fresh forms.

"Habitat with Date Stems."
Cast handmade paper;
cedar bark, abaca,
date palm stems, salt cedar,
goat vertebrae,
waxed linen, dye.
22" high x 10" wide x 10" deep.

Photo by Margot Geist

"Oracles." Cast handmade paper; oak leaf, abaca, salt cedar, goat and phython bones, porcupine quill, waxed linen, dye. 24" high x 12" wide x 12" deep.

Jackie Abrams

Jackie Abrams's work is about the possibilities of basketry—the possibilities of materials, of techniques, of colors and textures. It is a constant learning process, a progression of explorations. She is intrigued by the combinations of materials and techniques, the layers in a basket, and the painted surfaces that look like stone, leather, or ceramics. She is strongly inspired by architecture, other cultures, and the world around her. The challenge to keep exploring satisfies her sense of creative wonder.

Abrams's roots in basketry are strongly grounded in tradition. In 1975, she apprenticed to Ben Higgins, an 81-year-old basketmaker who worked with white ash. For 13 years, she wove traditional, functional baskets of natural materials. In 1984 she started to further explore the aesthetic possibilities of basketry. She has been working with archival cotton paper, an extremely versatile and user-friendly material, since 1990.

Recently, Abrams's work has been included in *Beautiful Things; Baskets—A Book for Makers and Collectors*; *Baskets: Tradition and Beyond*; *Fiberarts Design Book Six*; and *Making the New Baskets*. This year, shows include the ACC Baltimore Show, a three-person exhibit at the Works Gallery in Philadelphia in June, and basket exhibits at both the delMano Gallery in Los Angeles and the Dane Gallery in Nantucket. The Katie Gingrass Gallery will represent her at both SOFA New York and Chicago.

Photo by Jeff Baird

"Green Lines and Wire."
A woven basket of cotton paper,
acrylic paint, copper wire, varnish.
A study in form, color, and surface texture.
12" high x 4" wide x 5" deep.

100

"Copper Images." Stitched panels of cotton paper, copper, acrylic paint, wire, waxed linen, varnish. 11" high x 7" wide x 3" deep.

Linda Gunn

Influenced by her grandfather, one of the first Disney animators, and a mother who exposed her to the arts through theater, ballet, literature, and museum visits, Linda Gunn's interest in art began at an early age. Most significantly, a gift of John Gnagy's Art Studio in a Box—a set containing oil pants, watercolors, charcoal, pastel, and various drawing supplies, along with step-by-step instructions for each medium—helped structure Gunn's art education and focus her artistic goals.

"I have become a professional artist simply through the acts of observation, drawing, painting, and teaching," explains Gunn. "I paint from personal life experiences."

Educated at Long Beach City College, Gunn's career as an instructor came about unexpectedly, when she was a finalist in *The Artist's Magazine* Floral Painting Competition. After the image was published in the magazine and featured on the cover of a *Northlight* book, she was invited to teach watercolor classes for the Long Beach Recreation Department.

An avid traveler, Gunn maintains a journal/sketch book to record observations, new experiences, and references. These, combined with photographs, aid her in adding mood to her inspiring paintings.

*"Moo For Pam." Tinted collage papers applied over a simple sketch.
White gouache added for detailing; value changes and shapes outlined with inks.
11" high x 15" wide. Collection of Pam McKie.*

In 1995, Gunn's dream was to become a member of The National Acrylic Painters Association of Great Britain. After reviewing her portfolio at an emergency board meeting, NAPA invited her to join. Gunn became their first international member, setting the stage for the birth of NAPA USA, of which she is founding director.

Ensconced in her studio, Gunn works with all manner of media that can be mixed with water. Learning about the properties of acrylic, gouache, egg tempera, dyes, inks, and transparent watercolor has helped her to discover which medium will produce the best result for the desired effect.

In addition to NAPA, Gunn maintains affiliations with the National Association of Women Artists, International Society of Experimental Artists, and Women Painters West. She has been featured in such publications as *The Artist's Magazine; Watercolor Magic; Southwest Art; The International Artists Magazine;* and Russia's first art instruction magazine, *X.CoBet* (the cover and feature article of the first issue). She is also included in the books *The Artistic Touch* and *The Best of Watercolor: Painting Color.* Most recently, Gunn has been listed in the *Who's Who of American Women, 2000–2001* and has three new magazine articles being published.

"L.A. Zoo." Various textured papers that have been painted with watercolor, then applied as a mosaic over a completed painting. The elephants were collaged over with colored papers. White gouache was added for highlights and shapes were outlined with black ink. 30" high x 45" wide. Gallery Ishikawa, Yokohama, Japan.

"The Greeting." Collage papers that have been painted with watercolor,
then applied as a mosaic over a completed painting. Value contrasts with watercolor
and gouache; value changes and shapes outlined with inks.
15" high x 22" wide. Collection of M. Graham & Co., Inc.

"Point Fermin Lighthouse." Collage papers that have been painted with watercolor, then applied
as a mosaic over a completed painting. Value contrasts with watercolor and gouache;
paint splattered over entire composition; value changes and shapes outlined with inks.
15" high x 22" wide.

"Mountain Wind." Colored pulps processed from mulberry inner bark are poured onto a screen to form a base layer, then overlayed with additional colored pulps to create the background landscape. Strips of mulberry inner bark are laid into the background to make the tree structure. Once pulp is dry, it is peeled from the screen—a finished art piece. 5½' high x 3' wide.

Susan Olsen

In 1973, Susan Olsen learned the rare technique of "the paper is a painting" with a master in Obara Village, a small papermaking hamlet in Japan. Colored fiber pulps are processed from plant structures. The structures are cooked in soda ash to remove the plant glue from the cellulose. The fiber bundles are then bleached and dyed and macerated into liquid pulp. Containers of these colored fiber liquids make up the "palette." They are applied onto a screen and shaped into images. When the pulp layer dries, it becomes paper, is peeled from the screen, and . . . "the paper is a painting!"

Olsen's subjects express a deep desire within to see positive changes take place on Earth. Over the years she has been developing four main bodies of work that express this motivation: *Trees & Landscapes* are odes to the beauty and perfection of the natural world; *Primitive Images* are inspired by symbols of primitive peoples that tell of their innate knowingness of the connectedness of all being; *Process Abstractions* inspired by a macro view of healing occuring in the nervous system; and *Portraits of Their Essence*, which are portraits of beloved pets, primarily dogs.

Olsen's art pieces range in size from small to 5' x 9'. She also creates lamps, translucent screens, and paper fresco—where the paper is applied directly to the walls of a space.

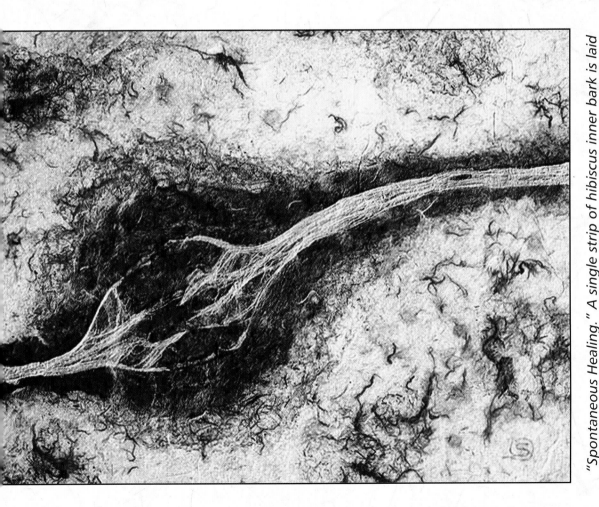

"Spontaneous Healing." A single strip of hibiscus inner bark is laid onto a screen to make the nerve structure. Colored pulps processed from mulberry inner bark and banana leaf stem are added, and then a layer is poured to cover the entire screen. The front of the art piece is that which faces the screen after it has been peeled off. 24" high x 36" wide.

Cleo Campos Hill

Cleo Campos Hill is a Santa Fe artisan, who makes hand-decorated lamp shades, using handmade oriental papers of many different colors and textures. Hill was always surrounded by color and light in her native Brazil. This gave her a love of playing with color and light in her compositions during the fifteen years she spent as an active artist in Sao Paulo. Hill apprenticed for a well-known Brazilian master papermarker for ten of those years. During this apprenticeship she learned every step in the art of papermaking and how to use paper in artistic installations and sculptures. After moving to New Mexico, she used her talents to start Shades by Cleo as a creative expression of color in functional art.

Hill's husband, Solomon, began making lamp bases to complement Hill's lamp shades. His inspiration is drawn from the beauty of the natural world. His love of the desert can be seen in his choice of materials that are common to the desert landscape. The marriage of these colorful materials with the colors and textures of Hill's paper lamp shades creates truly unique and aesthetically pleasing lamps.

Through their combined efforts, they bring you these beautiful pieces of functional art in the hope that you enjoy them as much as they do.

"Nestor."
Wall shade made from a collage
of Nepalese handmade papers.
15" high x 8" wide

"Ying-Yang." End table lamp with lamp shade made from Japanese momi papers.
Base made of granite and salt cedar twigs.
15" high x 23" wide

"Mint Green Lamp Shade." End table lamp with lamp shade made from mulberry and mango papers. Base made of lichen-covered sandstone.
15" high x 23" wide

Metric equivalency chart

INCHES TO MILLIMETRES AND CENTIMETRES

MM-Millimetres **CM-Centimetres**

INCHES	MM	CM	INCHES	CM	INCHES	CM
$1/8$	3	0.9	9	22.9	30	76.2
$1/4$	6	0.6	10	25.4	31	78.7
$3/8$	10	1.0	11	27.9	32	81.3
$1/2$	13	1.3	12	30.5	33	83.8
$5/8$	16	1.6	13	33.0	34	86.4
$3/4$	19	1.9	14	35.6	35	88.9
$7/8$	22	2.2	15	38.1	36	91.4
1	25	2.5	16	40.6	37	94.0
$1 1/4$	32	3.2	17	43.2	38	96.5
$1 1/2$	38	3.8	18	45.7	39	99.1
$1 3/4$	44	4.4	19	48.3	40	101.6
2	51	5.1	20	50.8	41	104.1
$2 1/2$	64	6.4	21	53.3	42	106.7
3	76	7.6	22	55.9	43	109.2
$3 1/2$	89	8.9	23	58.4	44	111.8
4	102	10.2	24	61.0	45	114.3
$4 1/2$	114	11.4	25	63.5	46	116.8
5	127	12.7	26	66.0	47	119.4
6	152	15.2	27	68.6	48	121.9
7	178	17.8	28	71.1	49	124.5
8	203	20.3	29	73.7	50	127.0

Acknowledgments

Jackie Abrams
21 Howard Street
Brattleboro, VT 05301
(802) 257-2688
www.jabrams@together.net

Colleen Barry
3715 Cal Ore Drive
Redding, CA 96001
(530) 243-6851

Barbara Fletcher
17 Power House Street #318
S. Boston, MA 02127
(617) 268-8644
www.paperdimensions.com/
frames.htm

Linda Gunn
2525 E. 5th Street
Long Beach, CA 90814
(562) 439-3276
www.AcrylicPainters@aol.com

Cleo Campos Hill
40 Likely Road
Santa Fe, NM 87505
(505) 986-0309
www.shadesbycleo.com

Susan Olsen
144 Hailey Drive
Franklin, NC 28734
(828) 524-7438
www.susano@smnet.net

Karen Simmons
P.O. Box 1161
Cedar Crest, NM 87008
(505) 281-1458
www.jaskaren@aol.com

We would like to thank Dreamweaver Stencils for providing the brass embossing stencil used for the Celtic knot design on Technique 7: Dry-embossed Card on page 42 and Beedz/Art Accents for providing the micro glass beads for the Carnival Mask shown on page 88.

Index